NATIONAL ACADEMIES *Sciences Engineering Medicine*

NATIONAL ACADEMIES PRESS
Washington, DC

Enhancing Coordination and Collaboration Across the Land-Grant System

Committee on Enhancing Coordination Between Land-Grant Universities and Colleges

Board on Agriculture and Natural Resources

Division on Earth and Life Studies

Board on Higher Education and Workforce

Policy and Global Affairs

Consensus Study Report

NATIONAL ACADEMIES PRESS 500 Fifth Street, NW Washington, DC 20001

This activity was supported by a contract between the National Academy of Sciences and the National Institute of Food and Agriculture in the U.S. Department of Agriculture. Any opinions, findings, conclusions, or recommendations expressed in this publication do not necessarily reflect the views of any organization or agency that provided support for the project.

International Standard Book Number-13: 978-0-309-69107-9
International Standard Book Number-10: 0-309-69107-9
Digital Object Identifier: https://doi.org/10.17226/26640

This publication is available from the National Academies Press, 500 Fifth Street, NW, Keck 360, Washington, DC 20001; (800) 624-6242 or (202) 334-3313; http://www.nap.edu.

Printed in the United States of America.

Suggested citation: National Academies of Sciences, Engineering, and Medicine. 2022. *Enhancing Coordination and Collaboration Across the Land-Grant System*. Washington, DC: The National Academies Press. https://doi.org/10.17226/26640.

The **National Academy of Sciences** was established in 1863 by an Act of Congress, signed by President Lincoln, as a private, nongovernmental institution to advise the nation on issues related to science and technology. Members are elected by their peers for outstanding contributions to research. Dr. Marcia McNutt is president.

The **National Academy of Engineering** was established in 1964 under the charter of the National Academy of Sciences to bring the practices of engineering to advising the nation. Members are elected by their peers for extraordinary contributions to engineering. Dr. John L. Anderson is president.

The **National Academy of Medicine** (formerly the Institute of Medicine) was established in 1970 under the charter of the National Academy of Sciences to advise the nation on medical and health issues. Members are elected by their peers for distinguished contributions to medicine and health. Dr. Victor J. Dzau is president.

The three Academies work together as the **National Academies of Sciences, Engineering, and Medicine** to provide independent, objective analysis and advice to the nation and conduct other activities to solve complex problems and inform public policy decisions. The National Academies also encourage education and research, recognize outstanding contributions to knowledge, and increase public understanding in matters of science, engineering, and medicine.

Learn more about the National Academies of Sciences, Engineering, and Medicine at **www.nationalacademies.org**.

Consensus Study Reports published by the National Academies of Sciences, Engineering, and Medicine document the evidence-based consensus on the study's statement of task by an authoring committee of experts. Reports typically include findings, conclusions, and recommendations based on information gathered by the committee and the committee's deliberations. Each report has been subjected to a rigorous and independent peer-review process and it represents the position of the National Academies on the statement of task.

Proceedings published by the National Academies of Sciences, Engineering, and Medicine chronicle the presentations and discussions at a workshop, symposium, or other event convened by the National Academies. The statements and opinions contained in proceedings are those of the participants and are not endorsed by other participants, the planning committee, or the National Academies.

Rapid Expert Consultations published by the National Academies of Sciences, Engineering, and Medicine are authored by subject-matter experts on narrowly focused topics that can be supported by a body of evidence. The discussions contained in rapid expert consultations are considered those of the authors and do not contain policy recommendations. Rapid expert consultations are reviewed by the institution before release.

For information about other products and activities of the National Academies, please visit www.nationalacademies.org/about/whatwedo.

COMMITTEE ON ENHANCING COORDINATION BETWEEN LAND-GRANT UNIVERSITIES AND COLLEGES

KATHRYN J. BOOR (*Chair*), Dean, Graduate School, and Vice Provost, Graduate Education, Cornell University
OLGA U. BOLDEN-TILLER, Dean, College of Agriculture, Environment and Nutrition Sciences, and 1890 Research Director, Tuskegee University
DINA J. CHACÓN-REITZEL, Executive Director, New Mexico Beef Council
MARTIN A. DRAPER, Associate Dean for Research, College of Agriculture; Director for Research, K-State Research and Extension, Kansas State University
RONNIE D. GREEN, Chancellor, University of Nebraska, Lincoln
MOSES T. KAIRO, Professor, Dean, and Director of Land-Grant Programs, School of Agricultural and Natural Sciences, University of Maryland Eastern Shore
JAN E. LEACH, University Distinguished Professor, Agricultural Biology Department; Associate Dean for Research, College of Agricultural Sciences, Colorado State University
KAREN PLAUT, Dean, College of Agriculture, and Professor of Animal Science, Purdue University
WENDY POWERS, Dean, College of Agricultural, Human, and Natural Resource Sciences, Washington State University
RYAN F. QUARLES, Kentucky Commissioner of Agriculture
HAROLD H. SCHMITZ, Senior Scholar, Graduate School of Management, University of California, Davis; General Partner, March Capital US, LLC
STEVE YANNI, Land Grant Director, Bay Mills Community College

Study Staff

ROBIN SCHOEN, Director, Board on Agriculture and Natural Resources
MARIA ORIA, Senior Program Officer (until December 2021)
SARAH KWON, Senior Program Assistant (until May 2022)
FRANCES E. SHARPLES, Science Writer
PAIGE JACOBS, Program Assistant
SAMANTHA SISANACHANDENG, Program Assistant

Consultant

H. MICHAEL HARRINGTON, Western Association of Agricultural Experiment Station Directors (Retired)

Reviewers

This Consensus Study Report was reviewed in draft form by individuals chosen for their diverse perspectives and technical expertise. The purpose of this independent review is to provide candid and critical comments that will assist the National Academies of Sciences, Engineering, and Medicine in making each published report as sound as possible and to ensure that it meets the institutional standards for quality, objectivity, evidence, and responsiveness to the study charge. The review comments and draft manuscript remain confidential to protect the integrity of the deliberative process.

We thank the following individuals for their review of this report:

NANCY COX, University of Kentucky
LEVON ESTERS, Purdue Polytechnic Institute
ROLANDO FLORES GALARZA, New Mexico State University
KENRETT JEFFERSON-MOORE, North Carolina Agricultural and Technical State University
AMBER MARLOW, Lac Courte Oreilles Ojibwe University
ORLANDO MCMEANS, Southern University and A&M College
RACHEL MELNICK, Agriculture & Food Systems Institute
DONALD R. ORT, University of Illinois at Urbana-Champaign
MARC PARLANGE, University of Rhode Island

Although the reviewers listed above provided many constructive comments and suggestions, they were not asked to endorse the conclusions or recommendations of this report, nor did they see the final draft before its release. The review of this report was overseen by **CUTBERTO GARZA,** Cornell University, Emeritus, and **NORMAN SCOTT,** Cornell University, Emeritus. They were responsible for making certain that an independent examination of this report was carried out in accordance with the standards of the National Academies and that all review comments were carefully considered. Responsibility for the final content rests entirely with the authoring committee and the National Academies.

Preface

The land-grant system was created in 1862 when President Abraham Lincoln signed the first Morrill Act into law. At the time, the Act committed a total of more than 17 million acres of federally controlled land to support establishment of colleges and universities that promised to provide education in practical subjects (e.g., mechanical arts and agriculture) in the context of a liberal education. Through a series of federal acts, land-grant colleges and universities have a three-part mission in the agricultural sciences: research, education, and extension. Further legislation contributed to the expansion of the system to include Historically Black Colleges and Universities (1890) and Tribal Colleges and Universities (1994). Today, at a time when science and technology are more important than ever, the system is a crucial asset for education and research for the nation's agricultural industries. However, the widespread geographic distribution of the land-grant colleges and universities coupled with their range of sizes and widely differing histories contribute to a system that is not comprehensively cohesive.

Meanwhile, the benefits and importance of multidisciplinary and collaborative research are clearly well recognized. Collaboration among practitioners in the life sciences, engineering, computing and data analytics, and the social sciences among other disciplines is furnishing new pathways to advances and innovations in science and technology. For agriculture, these new capabilities are potential game changers that are becoming essential to counter the impacts of changes in the world's climate, land use, human population size, plant and livestock pests and diseases, and other factors that threaten the health and welfare of the Earth's population and its natural resources.

Forward-thinking members of Congress have recognized the importance of these opportunities and have asked the National Academies of Sciences, Engineering, and Medicine to assess the key factors that make for successful collaborative science and what actions might be taken to enhance the impact of inter-institutional activities specifically across the land-grant system. In response, the National Academies established the Blue Ribbon Panel that I have had the pleasure of chairing and that has produced this brief report. This "fast-track study" was on an accelerated schedule that meant that the work of the Panel is only a beginning. A request for a follow-on project has already been received by the National Academies, and it is my hope that congressional interest in improving and promoting greater integration and collaboration among the land-grant colleges and universities with a consequent increase in the impact of agricultural and food science advances will remain high.

The development of this report would not have been possible without the dedicated work of the National Academies' staff. The team was ably led by Robin Schoen, the director of the Board on Agriculture and Natural Resources. Fran Sharples assisted with the writing of the report. H. Michael Harrington, former Executive Director of the Western Association of Agricultural Experiment Station Directors, served as a consultant and lent the Panel his in-depth expertise in agricultural extension. Paige Jacobs and Samantha Sisanachandeng provided crucial administrative support as the program assistants for the project. Maria Oria served as a senior program officer on the project until December 2021, and Sarah Kwon was its senior program assistant until May 2022. The Panel is grateful to all of these individuals.

Kathryn J. Boor, *Chair*
Committee on Enhancing Coordination Between
Land-Grant Universities and Colleges

Contents

Acronyms and Abbreviations

AFRI	Agriculture and Food Research Initiative
AI	artificial intelligence
ARS	Agricultural Research Service
CAP	Coordinated Agricultural Project
COA	College of Agriculture
COE	Center of Excellence
ESCOP	Experiment Station Committee on Organization and Policy
GAO	U.S. Government Accountability Office
HBCU	Historically Black College and University
IPM	integrated pest management
LTAR	Long-Term Agroecosystem Research
LTER	Long-Term Ecological Research
MAES	Montana Agricultural Experiment Station
MILES	Michigan Inter-Tribal Land Grant Extension System
NARETPA	National Agricultural Research, Extension, and Teaching Policy Act
NIFA	National Institute of Food and Agriculture
NIH	National Institutes of Health
NIMSS	National Information Management and Support System
NIPMCC	National IPM Coordinating Committee
NSF	National Science Foundation
OWCAP	Ogallala Water Coordinated Agriculture Project
SAES	State Agricultural Experiment Stations
SAS	Sustainable Agricultural Systems
SCRI	Specialty Crop Research Initiative
TCU	Tribal College and University
USDA	U.S. Department of Agriculture

Summary

This report examines the potential for land-grant colleges and universities to increase the impact of their collective contributions to the American public through inter-institutional coordination and collaboration. Established in 1862 by the first Morrill Act, the land-grant system began the democratization of post-secondary education in agriculture and other subjects across the United States, launching an initial set of academic institutions known as the "1862s." Subsequent legislation created the "1890s," which are Historically Black Colleges and Universities (HBCUs) that have agricultural and allied programs, and the "1994s," which are Tribal Colleges and Universities (TCUs). Today, there are 111 land-grant colleges and universities across all states and many U.S. territories. Through the U.S. Department of Agriculture's (USDA's) National Institute of Food and Agriculture (NIFA), the federal government invests money to carry out the system's tripartite mission of agricultural research, education, and extension, distributing funds to the states as both capacity grants and competitive grants to support land-grant colleges and universities. A wide distribution of academic institutions, together with associated experiment stations and extension services, work in partnership with states and counties to address a variety of local, state, and regional issues.

A congressional directive in the Consolidated Appropriations Act, 2021 called on USDA to establish a Blue Ribbon Panel to examine how cooperation in the land-grant system could deepen and expand the impact of its work. NIFA turned to the National Academies of Sciences, Engineering, and Medicine (the National Academies) for help in addressing the directive. To that end, the National Academies established the Committee on Enhancing Collaboration Between Land-Grant Universities and Colleges (the Panel) to examine how knowledge generation, problem solving, and opportunity creation across the food and agricultural knowledge system can be increased by enhanced collaborative activity.

The Panel's study was conducted in a compressed time frame over approximately 9 months, holding its first virtual public meeting in November 2021 and final virtual mini-workshop in August 2022. The Panel members met in closed virtual meetings to consider their charge, share individual experiences with collaboration, identify a variety of activities that might be broadly representative of the types of collaboration taking place across the land-grant system, and deliberate on the meaning of "success" for collaborative activities and the factors that contribute to success. It further explored the drivers of and potential benefits to expanding collaboration across the system, as well as perceived barriers to collaboration. These discussions led the Panel to develop a set of preliminary observations (see Appendix C) that were posted online for several weeks to obtain stakeholder comment. The feedback received (see Appendix D) prompted the Panel to organize virtual mini-workshops on several additional topics before drafting its report.

Given the explicit language of the congressional directive, the Panel's report focuses primarily on the dynamics of cooperation *inside* the land-grant system. It is recognized, however, that the land-grant system exists in a universe that includes many other non-land-grant colleges and universities that also conduct research and provide education, as well as private-sector entities that often collaborate with land-grant institutions. These relationships are relevant to broadening the impact of the land-grant system and should be included in future explorations of this topic.

COLLABORATION IN THE LAND-GRANT SYSTEM

The Panel found that collaborative projects between faculty members at different land-grant colleges and universities occur frequently. For example, there are currently 247 active multistate projects supported through legislatively appropriated funding (Hatch Act of 1887) directed to 1862 universities for the purpose of interstate coordination and collaboration on research issues of mutual interest and concern. Multistate research projects are initiated by the agricultural experiment stations in two or more states. Once established, they are open to other institutions, including other land-grant and non-land-grant colleges and universities, federal agencies, and the private sector, which bring their own sources of funding to participate. Some multistate projects are long-standing, decades-old activities. Others form to address a problem and eventually become inactive. However, few if any involve participants from the 1890 or 1994 institutions.

From examples of past and current collaborative projects, Panel members identified examples of projects with demonstrated "success." These included the large award-winning multistate project, Soil, Water, and Environmental Physics to Sustain Agriculture and Natural Resources, as well as the competitively funded Specialty Crop Research Initiative project on zebra chip disease of potatoes. Across and beyond the system, there are diverse kinds of projects, such as the large-scale, multi-institutional Ogallala Water Coordinated Agriculture Project funded through the Agriculture and Food Research Initiative (AFRI) competitive grants program of NIFA, as well as the National Integrated Pest Management Coordinating Committee supported by the Association of Public and Land-Grant Universities.

In general, there was a dearth of participation by 1890 and 1994 institutions. In contrast, the Panel found that greater inter-institutional collaboration happens at the state level.

In Virginia, extension is a shared responsibility of Virginia Tech (an 1862) and Virginia State University (an 1890), and in Alabama, through the Alabama Agricultural Land Grant Alliance, statewide extension activities are jointly conducted by Auburn (an 1862) and Alabama A&M and Tuskegee University (both 1890s). It is worth noting, however, that creating collaborations inside state lines may be easier than those between different states, which may be more complex due to differences in local priorities and in legislative frameworks.

In some places, a high-level partnership agreement creates the foundation for building trust and familiarity as a prerequisite for collaboration. An example is the Michigan Inter-Tribal Land Grant Extension System (MILES), which reflects the commitment of the state land-grant colleges and universities (1862s and 1994s) to the success of each other, the tribes, and the collective benefit of the public in the state. The 1890 Centers of Excellence,[1] established at HBCUs to leverage disciplinary diversity among institutions, provide a unifying space for collaborative activity. These and other examples of successful past or ongoing multidisciplinary and/or multi-institutional projects and programs reflect the diverse kinds of collaborative research across the land-grant system.

FRAMING THE VALUE OF COLLABORATIVE AND MULTIDISCIPLINARY RESEARCH

As there are already many active collaborations in the land-grant system, the Panel reflected on why collaboration deserves additional attention. One compelling answer is that the focus and complexity of key questions in food and agricultural science are evolving in ways that would be enhanced by collaborative effort. The 2019 National Academies' report *Science Breakthroughs to Advance Food and Agricultural Research by 2030* notes that the serious problems facing agriculture today are unlike those in the past. The natural resources on which agriculture depends, such as water and fertile soils, are showing serious signs of stress due to changes in climate, land use, population growth, and other factors that affect the agricultural and food system. The *Science Breakthroughs* report defined the major goals for food and agricultural research in the next decade to include "(1) improving the efficiency of food and agricultural systems, (2) increasing the sustainability of agriculture, and (3) increasing the resiliency of agricultural systems to adapt to rapid changes and extreme conditions" (NASEM, 2019, p. 2).

Those goals are underpinned by broad research challenges that can be addressed most powerfully by insights from multiple scientific disciplines that promote "convergent" solutions, as described in two previous National Research Council reports, *A New Biology for the 21st Century* (NRC, 2009) and *Convergence: Facilitating Transdisciplinary Integration of Life Sciences, Physical Sciences, Engineering, and Beyond* (NRC, 2014). The latter defines convergence as "an approach to problem solving that cuts across disciplinary boundaries and integrates knowledge, tools, and ways of thinking from life and health sciences,

[1] See https://www.nifa.usda.gov/grants/about-programs/program-operational-areas/1890-land-grant-institutions-programs, accessed September 20, 2022.

physical, mathematical, and computational sciences, engineering disciplines, and beyond to form a comprehensive synthetic framework for tackling scientific and societal challenges that exist at the interfaces of multiple fields" (NRC, 2014, p. 1).

The *Science Breakthroughs* report encouraged the use of novel, crosscutting tools such as the following to "break through" the difficult problems facing agriculture:

- Employing systems science to develop holistic understanding of problems;
- Developing field-deployable sensors for rapid and dynamic monitoring of conditions of interest across multiple scales and geographies;
- Using data science, software tools, and advanced analytic models to increase predictive abilities;
- Gene editing of agriculturally important organisms for productivity, quality, and climate resilience; and
- Harnessing the soil, plant, and animal microbiome to improve crop production, transform feed efficiency, and increase resilience to stress and disease.

This toolbox cuts across the biological sciences, engineering and technology, the human sciences, and economics, and reflects the need for convergent thinking. In a collaborative setting, the use of these applications will be more impactful with regional or national coordination and planning. The land-grant system, with its nationwide remit, seems an ideal place to implement recommendations from the *Science Breakthroughs* report.

Applications of Data Science in Agriculture

One of the more compelling collaborations the Panel learned about was the NIFA and National Science Foundation–funded Artificial Intelligence (AI) Institutes, which provide data science and AI tools to a group of diverse partners, including 1862 and 1890 institutions as well the USDA Agricultural Research Service, the Argonne National Laboratory, and the Donald Danforth Plant Science Center. In addition to the flexibility of the platform to accommodate different types of research, the platform strongly supports teaching and extension elements. This kind of collaborative platform may not be burdened by some of the problems that other collaborations face, such as bureaucracy and inequity in partnership. The potential to create different types of these platforms needs further exploration.

The Science of Team Science

Another ingredient to successful collaboration are the human factors involved, because even with cutting-edge science, a diversity of collaborators must work together. The science of team science offers provocative insights from the study of multidisciplinary and multisector science teams and can even provide teams with feedback on their connectedness during the course of a collaboration. Jennifer Cross of Colorado State University, who presented her research to the Panel, said the data show there is no shortcut around the development of trust and a shared vision if a collaboration is to be successful, but these things can be built proactively. Team science has a theory, and it can be predictive. Such information could be useful for any scientific team.

Capacity Funding for the Land-Grant System

Without a doubt, collaboration requires sufficient funding. Unlike the legislative authorities (Hatch funds) that support multistate collaboration in the 1862 institutions, no such designated pot of money exists for the 1890s and 1994s. Other funds could be used, but these institutions are already underfunded in general. Many are not able to raise the 1:1 match required to receive full federal funding. In 2020, 10 of the 19 1890 institutions did not meet the match and collectively lost out on $21 million. The inequity of the funding has prompted lawsuits in some states, most recently Tennessee and Maryland, to challenge the legacy of providing state-matching funds at states' majority-White universities but not states' HBCUs (1890s). Most of the funds to the 1994 institutions (TCUs) comes from interest off a public endowment.

An alternate to capacity funding could be the competitively funded grants of AFRI. In 2021, an HBCU (Central State University) was awarded a grant to lead a Coordinated Agricultural Project grant with 1862 and 1994 institutions. Success rates of AFRI grants range between 10 and 40 percent. A larger backdrop to the funding picture is that state and federal funding for agricultural research has been declining for two decades, dropping from a peak of $7.64 billion in 2002 (in 2019 dollars) to $5.15 billion in 2019, the same level as in 1970. Considering the return on every dollar spent is conservatively valued at $20, U.S. competitiveness is being short-changed.

PRINCIPLES FOR ENHANCING THE SUCCESS AND IMPACT OF LAND-GRANT COLLABORATION

The Statement of Task asks the Panel for a framework of principles to foster successful coordination and collaboration in the land-grant system. The Panel's view is that effective systems-based approaches will require the following:

- Support for large-scale collaborative projects at the highest administrative levels in the various institutions;
- Targeted financial resources to enable planning and communication among institutions;
- Focused operating resources, including administrative support and trained project and data management support, in addition to directed funds to support research, teaching, and outreach;
- Incentive structures that enable faculty to participate in large-scale projects without compromising on their other responsibilities;
- Uniform, shared data management systems that enable seamless access to emerging information; and
- Outstanding communications support to inform the public, including legislators, of the outcomes of their financial investments.

RECOMMENDATIONS TO OVERCOME BARRIERS TO COLLABORATION

The Panel was also asked to make recommendations to overcome barriers and offers these five that could be impactful in improving the success and impact of collaboration in the land-grant system:

1. **It is urgent that Congress take action to facilitate the participation of all land-grant colleges and universities in multistate research and extension projects.**

Currently, inter-institutional collaborations of the multistate research projects do not routinely engage faculty from the full range of institutions across the land-grant system. Collaborations that involve the 1890s and 1994s occur much less often than collaborations among the older and better funded 1862s. As noted earlier, different authorities guide the allocation of federal funds to 1862, 1890, and 1994 institutions respectively, with different requirements for state-matching support of federal dollars and different stipulations for the use of funds for collaboration. Historical and current funding disparities have prevented many 1890s and 1994s from being full partners in collaborations with the 1862s. If unifying the land-grant system around common national pursuits is important, Congress should address the need to provide dedicated funding to support participation in collaborative activities.

2. Land-grant administrators should examine how to reduce the variability in committed support for faculty participation in collaborative activities.

Institutions have differing approaches for supporting faculty involvement in collaborative research. That creates varying expectations about the nature of the activity and the role of participants. For example, some institutions use multistate research funds to support salaries while others may use them to support travel to participate in meetings. The Panel recognizes that experiment station and extension directors have reasons for managing their funds as they do, but expectations for the level of participation in a collaborative activity need to be made clearer by parties involved along with decision making around values.

3. Faculty members in academic departments should reflect on how collaborative activities fit into an academic career and advocate for their reward.

Like many, if not most, academic entities, land-grant colleges and universities have traditions emphasizing and rewarding competitive, rather than collaborative, research projects. A stakeholder commented that "departmental cultures around collaborations vary widely but are constantly changing. In the natural sciences, any institution that has failed to keep up with the collaborative nature of research in their evaluation procedures will fail to be competitive at all." If this is true, then the time has come for departments to modify evaluation criteria for promotion and tenure to ensure that achievements in team science are appropriately recognized and rewarded.

4. Funding agencies should help faculty find partners for collaboration.

Several respondents from 1862 institutions to the Panel's preliminary observations acknowledged the problematic nature of not involving participants from the 1890 and 1994 institutions and even expressed dismay at being unaware of the expertise at institutions across the land-grant system, not to mention at non-land-grant colleges and universities or other organizations that may serve as suitable partners for collaboration.

To increase knowledge of where relevant expertise exists across the land-grant system, NIFA could encourage collaboration by convening information exchange or sandbox workshops. These could highlight funded research projects or new areas of research and allow investigators to share information about work they may be doing in topical areas that are of interest to other scientists in the system. This could, in turn, help to establish relationships among investigators with common interests in particular problems and, as suggested above, help create interpersonal foundations for future collaborative work.

Several individuals from 1994 institutions emphasized the need to build trust and relationships prior to inviting them to be included on a proposal or even to participate in meetings, which makes yet another claim on limited time. The collaborative activities between Tribal Councils and the State Agricultural Experiment Station and Extension in Montana as well as MILES (see Chapter 2) may be useful models for partnering with 1994 institutions.

5. Land-grant colleges and universities should help build capacity for collaboration by introducing faculty to the science of team science.

Establishing collaborations requires team building, emotional intelligence, and project management skills, for which many faculty are unprepared and untrained. Academic institutions generally only aim to train students in their scientific disciplines, not in understanding the human behaviors that foster successful teams. There is a scientific basis for understanding these dynamics in collaboration, and academic institutions should use that knowledge to inspire more explicit thinking around team dynamics and self-awareness by leaders of and participants in collaborative activities. One respondent to the Panel's preliminary observations pleaded that the Panel does not recommend "another useless training exercise." Institutions should pilot ways to expose faculty to the science of team science so that potential for skill-building is supported by credible feedback.

RECOMMENDATIONS FOR AMPLIFYING IMPACTS THROUGH EFFECTIVE COMMUNICATIONS

The Panel offers two recommendations for improving public awareness of the work of the land-grant system:

1. **NIFA should support a collaborative research and extension activity of social scientists and science communicators in the land-grant system to develop outreach strategies.**

Documenting the economic, environmental, and societal impacts and benefits of collaborative research is a way of raising the profile of this information for producers, policy makers, and the public. Simply stated, impact is the effect that an activity or project, especially something new, has on a situation, person, or policy. Constituents may be more likely to recognize the impacts and benefits of collaborative activities among land-grant colleges and universities if communication strategies are tailored to their interests and levels of understanding.

2. **Land-grant colleges and universities should create novel messaging vehicles to reach specific audiences about the outcomes of collaborations.**

Effectiveness can be enhanced by creating specific messages for particular audiences and by using tools that increase audience understanding. For example, the use of graphical tools can be an effective method of communication for some audiences. Telling the story behind the research, perhaps by using analogies, and clearly explaining why the results are important for ordinary citizens or policy makers could be helpful for nonprofessional audiences. Student-led hackathons could bring a fresh approach to information dissemination.

NEXT STEPS

The Panel notes that the principles and recommendations in its report provide only an initial framework to enhance collaboration across the land-grant system that would allow it to operate more effectively as a unified whole. Adopting a culture of collaboration implies change in behavior reinforced by enabling policies. For example, faculty would be encouraged to pursue collaboration more readily if the time required for project planning and team development and maintenance were treated as valuable during tenure and promotion deliberations. The adoption of curricula for project management, communication, and other skills that enhance collaboration would improve chances for team success and perhaps provide new and satisfying career pathways for students and faculty members.

Diverse backgrounds, cultures, community relationships, and scientific expertise reside in the land-grant colleges and universities, but collaborative vehicles are needed to take advantage of that diversity and allow the institutions to operate in a more connected way. Congress can play a role in supporting their development through examining its support of the land-grant colleges and universities and the statutes that encourage collaboration. Some of the obstacles mentioned in the report are not trivial to overcome. Addressing some of them implies the need for additional funding and responsibility in return for greater effectiveness, capacity, and "true" partnership. The ability to reach stakeholders and the public more effectively may be pivotal to the ability to get the support needed to achieve greater collaborative outcomes.

The Panel hopes that the ideas offered in this report will gain the attention of leaders of the land-grant colleges and universities and their stakeholders, who will then be willing to spend time developing and implementing them with the support of Congress.

1

Introduction

In 1862, Congress created the land-grant system through the first Morrill Act. The Act established a new policy of providing federal support for postsecondary education in agriculture, the "mechanical arts," and other subjects, specifically enabling citizens across the socioeconomic spectrum to access educational opportunities previously available only to the "landed gentry" (CRS, 2019). In total, 111 land-grant colleges and universities have been established across all states and many territories, including American Samoa, Guam, Micronesia, the Northern Marianas, Puerto Rico, and the U.S. Virgin Islands.

The Morrill Acts of 1862 and 1890 and the Equity in Educational Land-Grant Status Act of 1994 created three institutional categories of the land-grant system, now known as the "1862s," "1890s," and "1994s." The 1862 institutions were the first land-grant colleges and universities; the 1890 institutions are Historically Black Colleges and Universities (HBCUs) with agricultural and allied programs; and the 1994 institutions are Tribal Colleges and Universities (TCUs). Subsequent legislation recognized additional categories for specific programs, including non-land-grant colleges of agriculture and Hispanic-serving agricultural colleges and universities.

The national system of land-grant colleges and universities created under this legislative framework has been noted for its breadth, reach, and excellence in research, education, and extension. Kopp (2021) wrote that the "land-grant universities, together with the associated agricultural experiment stations and cooperative extension services, have played a crucial role in democratizing scientific knowledge and addressing intertwined educational, environmental, economic, and democratic challenges within the USA. Indeed, they have arguably pioneered the idea of 'usable science.'"

Currently, the identity of each land-grant entity is one of an independent enterprise rather than as a member of a national, integrated, and comprehensive system. In large part, each is dedicated to serving the clientele of its home state by providing opportunities for higher education, basic and applied research, and extension services. At the time of the first Morrill Act in 1862, that meant meeting the needs of the farming communities to which most Americans belonged. The agricultural industry remains an important focus of many land-grant colleges and universities, even as their portfolios are evolving to serve more diverse interests and needs. As an essential societal endeavor, agriculture in the United States has undergone tremendous change in the past 150 years, and the scientific and practical understanding of what makes it successful and sustainable continues to evolve. The production of food, fiber, and fuel is a complex biophysical and socioeconomic activity built on a finite natural resource base and shaped by (increasingly) variable abiotic and biotic environmental forces and by the effects of science and technology, input and output supply chains, consumer needs, policy, and regulation. No single institution can provide all of the answers to the questions that today's agricultural producers must address, much less satisfy all that society wants from the food and agriculture system.

For this reason, the co-development and support of collaborative, systems-based approaches in research, education, and outreach across multiple land-grant colleges and universities to address the multifaceted problems now facing the agricultural and food systems are urgent needs that have been recognized by policy makers (e.g., CRS, 2019). A congressional directive in the Consolidated Appropriations Act, 2021 called on the U.S. Department of Agriculture (USDA) to establish a Blue Ribbon Panel to examine how cooperation in the land-grant system "including the 1890 institutions" could deepen and expand the impact of its contribution to meeting national needs and global food security.

The National Institute of Food and Agriculture (NIFA) turned to the National Academies of Sciences, Engineering, and Medicine (the National Academies) for help in addressing the directive. To that end, the

National Academies established the Committee on Enhancing Collaboration Between Land-Grant Universities and Colleges (the Panel) to explore the potential to enhance the inter-institutional collaboration of participants from a diversity of land-grant colleges and universities to increase knowledge generation, solve problems, and create positive opportunities across the food and agricultural knowledge system. Increasing collaboration across the land-grant system would also improve both the perception and the reality of the system as an integrated whole with common goals and purposes.

The National Academies appointed the Panel members on the basis of their experience and expertise, and to be collectively representative of the different kinds of institutions (1862s, 1890s, 1994s) and stakeholders in the land-grant system. The Panel members include university leaders holding a wide range of leadership positions within land-grant colleges and universities as well as a state commissioner of agriculture, the executive director of a state beef cattle council, a member of a state Board of Regents, a venture capitalist, and a former science officer for a major food company. (See biographies of the Panel members in Appendix A.)

To guide its deliberations, the Panel was given a Statement of Task, which included some elements of process to facilitate the expedited ("fast-track") development of a consensus report as follows.

STATEMENT OF TASK

An ad hoc study committee (Blue Ribbon Panel) will identify key factors for successful outcomes of coordinated and collaborative projects between colleges and universities in the Land-Grant system, including those involving Historically Black Colleges and Universities, and other institutions, which address national challenges and global food security. The committee will prepare a report recommending actions to enhance the success and impact of inter-institutional activities.

To accommodate a compressed study timeline, the committee will explore the opportunities and limitations of inter-institutional projects by focusing on case studies chosen from the portfolio of past Multistate Research and Extension Activities, Coordinated Agricultural Projects (CAPs) of the Agriculture and Food Research Initiative, and other regional initiatives involving research, education, and extension. The committee will use the selected case studies to develop a framework of principles for fostering successful coordination and collaborations in pursuit of diverse outcomes. Outcomes may include those related to addressing specific national priorities, advancing knowledge, building human resource capacity, supporting commercial innovation, producing economic, environmental, and social benefits, and/or increasing public engagement. The committee will also consider factors that contribute to administrative goals of efficiency, ease, and transparency in reporting and in documenting impacts within and across projects.

After deliberating on the case studies, the committee will present to stakeholders, at a multiday virtual workshop, a conceptual overview of its framework of principles for successful coordinated and collaborative activities. Based on feedback from the workshop participants, the committee will finalize and describe the principles in a short report that recommends actions to reduce barriers to effective coordination and to foster impactful collaboration. The committee's report also will recommend processes that Land-Grant schools can use to capture and share successes, outcomes, and impacts of joint projects.

STUDY PROCESS: HOW THE PANEL CONDUCTED ITS WORK

In November 2021, the fast-track study began with a virtual, public meeting with USDA officials, including the Deputy Under Secretary for Research, Education, and Economics and Director of NIFA, to gain insight into the agency's view of the charge to the Panel. These officials suggested that the Panel should be expansive in its thinking and ambitious in its recommendations for actions by Congress, potentially as input to Farm Bill discussions, and by NIFA and the institutions of the land-grant system.

Following the initial meeting, the Panel held a series of closed meetings during which members explored impressions of and experiences with collaborative activities in the system. The Panel received a background presentation on the Multistate Research Program from a consultant, a former executive director of the Western Association of Agricultural Experiment Station Directors.

The collective history and perspectives of the Panel members, many of whom have experienced land-grant colleges and universities as undergraduate, graduate, and postdoctoral students, and later as faculty members and administrators, was the starting point for deliberations on what attributes of collaboration contribute to a definition of success. The Panel identified several examples of projects perceived by one or more members as having a reputation of success, but the Panel did not conduct an exhaustive review of all of the collaborative activities in the land-grant system. The examples selected by the Panel provided the basis for discussions around the diversity of activities and the limitations of collaboration.

Due to the tight project schedule and the lingering effects of the COVID-19 pandemic, it proved impossible to schedule a single, multiday workshop for stakeholders. Moreover, the Panel felt that there were additional ways to obtain feedback from stakeholders, including from those less likely to have an opportunity to comment during a workshop. Therefore, in the spring of 2022, the Panel developed 17 preliminary observations (see Appendix C) about collaboration in the land-grant system that were posted on the study website to solicit feedback from stakeholders. The website and portal for submitting comments on the preliminary observations were publicized in the e-newsletter of the National Academies' Board on Agriculture and Natural Resources and other National Academies' newsletters. An announcement was posted on the NIFA website and circulated through NIFA e-news. The Panel chair gave zoom briefings to the Association of Public and Land-Grant Universities' Extension Committee on Organization and Policy and Experiment Station Committee on Organization and Policy, and Panel members used their professional networks in academe, the National Association of State Departments of Agriculture, and elsewhere to publicize the portal.

The website and comment portal were open for 3.5 weeks beginning in mid-April 2022. The Panel received responses from 78 individuals (see Appendix D), reflecting a diversity of respondents that included agricultural producers; experiment station directors; extension administrators; land-grant directors and coordinators from TCUs and HBCUs; and associate, assistant, and full professors from 1862, 1890, and 1994 institutions. The full text of the comments is posted in the public access file for the study.[1] These comments helped shape the conclusions and recommendations of the Panel.

The comments received also prompted the Panel to explore several topics more deeply (see Appendix B), beginning with a public webinar in June 2022 on the science of team science, followed by three virtual mini-workshops held between July and August 2022:

- *Enhancing Collaboration and Deepening Impact: Can Data Science and Artificial Intelligence (AI) Enable New Collaborative Platforms Between Diverse Land-Grant Institutions and Create More Impactful Outcomes?* This workshop addressed the power of data science and analytics in food and agriculture "to accelerate research, expand America's workforce, and transform the future of the system." The workshop also illuminated how the attributes of the "collaborative platform" approach can overcome barriers to equitable participation in more conventional collaborations.
- *Building and Sustaining a Culture of Collaboration Across the Land-Grant System.* The workshop explored different examples of federally supported collaborative agricultural and other programs and projects that provide foundations on which collaborations can build.
- *The Role of Capacity for Collaboration in the Land-Grant System.* The workshop speakers shared their views on what capacity means in terms of their land-grant colleges and universities and the factors that allow participation in collaborative activities. They described what could help them build capacity and where capacity is most needed.

Informed by stakeholder feedback and the webinar and workshops, the Panel completed its report, which was anonymously peer reviewed according to the National Academies' procedures before being revised and subsequently approved for public release.

---

[1] See www.nationalacademies.org, accessed September 20, 2022.

ORGANIZATION OF THE REPORT

Chapter 2 of the report describes the multistate research projects, CAPs, and other examples of collaborative projects, and explores their benefits, limitations, and associated outcomes. Chapter 3 lays out a broad framing of what is at stake and the main ingredients needed to take on national challenges and global food security: cutting-edge science from across all disciplines, a strong collaborative organization, and availability of funding to support the work. Chapter 4 describes the beginning of a framework around principles for enhancing the success and impact of collaboration, by examining barriers to collaboration and the need to build communication strategies. The Panel offers seven recommendations. Chapter 5 discusses next steps in what must be a continued conversation around collaboration in the land-grant system.

Appendix A contains biographies of the Panel members. Appendix B is a listing of the agendas for public webinars and workshops described above and information on accessing video recordings of the events. Appendix C contains the Panel's preliminary observations posted for online comment. Appendix D is a brief summary of the topics raised in response, and information about the backgrounds of respondents.

2

Collaboration in the Land-Grant System

The Statement of Task directed the Panel to use examples of multistate research and extension activities, Coordinated Agricultural Projects (CAPs), and other regional initiatives involving research, education, and extension to explore the opportunities and limitations of inter-institutional projects, recognizing that collaborative projects may have diverse outcomes. The Statement of Task notes that outcomes may include "those related to addressing specific national priorities, advancing knowledge, building human resource capacity, supporting commercial innovation, producing economic, environmental, and social benefits, and/or increasing public engagement."

This chapter describes various ways in which collaboration is occurring within the land-grant system and includes several examples of ongoing activities perceived by one or more Panel members as having outcomes such as those listed in the Statement of Task (which appear here in bold in the text describing the examples). The Panel provides these examples not with the objective of finding the "most successful" projects but as illustrative of what is most typical in terms of collaboration in the land-grant system today.

MULTISTATE PROJECTS

The Multistate Research Program is supported by federal Hatch Act funds that are provided to 1862 institutions and their affiliated State Agricultural Experiment Stations (SAES) to work together on pressing problems that concern two or more states. To create a Multistate Research Program project, at least two SAES directors or extension directors must agree that a pressing problem needs to be addressed. After review of the initial request by a regional review committee, a nationwide call for participation is sent to all SAES and extension directors, the U.S. Department of Agriculture's (USDA's) Agricultural Research Service (ARS), and other appropriate agencies and individuals. Interested participants then develop the full scope of the project, which is reviewed and approved by the relevant regional association for an initial 5-year period. Competitive renewals may extend the project beyond 5 years. The Panel found details for 247 currently active multistate research projects on the National Information Management and Support System (NIMSS),[1] which includes the project objectives, publications, and a list of participants and organizations. Many of these projects have participants from across the land-grant system as well as representatives from industry, government, and non-land-grant colleges and universities. Participants generally bring their own sources of support to participate in these projects. Very few of the projects have participants from 1890 institutions. The NIMSS website does not enable a search for specific institutions, but a look at the individual details of 50 projects in the southern region found a total of eight 1890 participants (out of hundreds from other institutions) and none from the 1994 institutions.

Since 2012, nontechnical summaries of the research outcomes and economic impacts of multistate research projects have been developed with the help of a communications team that coaches scientists on how to write an impact statement. These are housed at the Multistate Research Fund Impacts website.[2]

Soil, Water, and Environmental Physics to Sustain Agriculture and Natural Resources

Not all multistate research projects are large-scale collaborations, but the Soil, Water, and Environmental Physics to Sustain Agriculture and Natural Resources project is one that is. It addresses an issue of

[1] See nimss.org, accessed September 20, 2022.

[2] See www.mrfimpacts.org/impact-statements, accessed September 20, 2022.

national importance: healthy soils. Soils are critical natural resources that manage, store, and supply water and nutrients to crops. Across the United States, soils are being degraded, the result of human-caused factors from changing land use practices to climate change. The project's impact statement notes, "For 60 years, researchers from more than 24 State Agricultural Experiment Stations have worked together to better understand how water, energy, and nutrients move through and interact with soil." In **advancing knowledge**, the project has helped state and federal agencies develop new tools and best management practices that provide **economic, environmental, and social benefits** such as reducing flood risks, storing nuclear waste, predicting drought and wildfires, and addressing dust and mine runoff. Members of the project shared information and **publicly engaged** with concerned Navajo farmers after the Gold King Mine Spill in 2015.

The HYDRUS software, which was developed by the project, is a model to assess hydraulic, thermal, biogeochemical, microbial, and gaseous processes in soils. It is used widely in soil physics, hydrology, and environmental science courses around the world, **building human capacity** globally in these fields. The project also **supported commercial innovation**, leading to the design of new sensors, data loggers, and other products sold worldwide to farmers, scientists, and other users.

The Soil, Water, and Environmental Physics to Sustain Agriculture and Natural Resources project received the Excellence in Multistate Research Award in 2021. Of the more than 50 university participants in the project, none are associated with an 1890 or a 1994 institution.

COORDINATED AGRICULTURAL PROJECTS

CAPs are multimillion-dollar ($10 million) grants awarded competitively for a 5-year period through the Agriculture and Food Research Initiative program administered by the National Institute of Food and Agriculture (NIFA). The competition is open to all universities, colleges, nonprofits, and other research organizations, and the scope of the projects is intentionally multidisciplinary and systems oriented. The current programmatic emphasis for CAP grants is Sustainable Agricultural Systems (SAS), and proposals must address one or more specified long-term goals: sustainable agricultural intensification, agricultural climate adaptation, value-added innovation, and food and nutrition translation. Within each of these goals are thematic areas that projects should address and expected outcomes. Projects must "result in societal benefits, including promotion of rural prosperity and enhancement of quality of life for all those involved in food and agricultural value chains from production to utilization and consumption."[3] There is a two-step process for applying. A letter of intent must be sent, followed by a final proposal. To be considered for funding, projects must include research, education, and extension elements. The distribution of funds to participating institutions is made as sub-awards from the lead institution. In 2021, NIFA made 15 CAP grant awards under the SAS program. For the first time ever, a Historically Black College and University (HBCU) (Central State University) was awarded $10 million and will be the lead performing institution, with partners from 1890, 1994, and 1862 institutions on the topic of sustainable aquaculture production of high omega-3-containing fish using a novel feed additive (hemp).

Unlike the multistate projects, there is no information repository dedicated solely to CAPs, though information about a specific CAP can be searched in USDA's Current Research Information System. Most CAPs develop their own websites to spotlight the progress of the collaboration during the course of its 5-year span.

The Ogallala Water Coordinated Agriculture Project

The Ogallala Water Coordinated Agriculture Project (OWCAP) is a large transdisciplinary project addressing a large-scale regional problem of **national importance** to the agricultural supply. It is focused on developing and sharing **practical, science-supported information** relevant to best management practices for optimizing water use across the Ogallala region. The Ogallala aquifer is a vital natural resource in

[3] See https://www.nifa.usda.gov/grants/funding-opportunities/agriculture-food-research-initiative-sustainable-agricultural-systems, accessed September 27, 2022.

the western United States; it underlies parts of Texas, Oklahoma, Kansas, Nebraska, Colorado, and New Mexico. Groundwater pumped from the Ogallala aquifer has transformed this region from a dustbowl to an agricultural powerhouse, producing more than 30 percent of U.S. crops and livestock and significantly increasing domestic and international food supplies. However, extensive groundwater pumping has led to significant depletion of the aquifer and declining water quality in certain areas. In 2016, NIFA funded the OWCAP project **to address the multifaceted challenges** facing the aquifer region through interdisciplinary and cross-institutional research and outreach. The 70-member interdisciplinary team from 10 institutions and six states **catalyzed new understanding** to identify and promote management of the Ogallala resource and support the region's communities. None of the team members were from 1890 or 1994 institutions.

The team's work focused on multiple scales of water management, including individual producers; local and regional entities, such as groundwater management districts; and the broader multistate aquifer region. Using its integrated approach, the team developed new tools, approaches, and partnerships to inform future water-management research and outreach. **Novel outreach mechanisms** such as a producer-led irrigator certificate program (producers instructing other producers) and efforts to develop an **economic benefit** of marketing a brand of beef raised using water-conserving practices are associated with this project. The project built on existing efforts that had been in place since 2003 to monitor and plan for Ogallala aquifer declines using ARS funding. The project illustrates that infusions of supplemental funding are critical in new discoveries and can be particularly impactful when a foundational level of support is already addressing an issue.

NATIONAL AND REGIONAL PROGRAMS

The first of the Panel's preliminary observations shared with stakeholders stated the following: *There is a significant amount of active and successful inter-institutional collaboration and cooperation taking place in the land-grant system today.* There was generally broad agreement about that statement from respondents. Individuals who commented gave a list of other types of successful ongoing collaborative work, including Regional Extension and Research Centers, Rural Development Centers, projects funded by the Sustainable Agriculture and Research and Education program, and public–private projects funded by the Foundation for Food & Agriculture Research,[4] among others. Some are funded by an external group but have a purpose and work closely with federally funded researchers, such as the following example.

National Integrated Pest Management Coordinating Committee

The National Integrated Pest Management (IPM) Coordinating Committee (NIPMCC) serves as an example of a project to address **the need for improved information exchange** among research and extension programs and the actions that were taken to address this problem. In approximately 2012, the IPM community recognized that much of the work it was doing, especially in the Cooperative Extension context, was occurring in a vacuum. Administrators across the system did not know what was being accomplished, nor did scientists in various other programs that had focused missions aimed at specific IPM audiences.

At the behest of IPM practitioners, the Experiment Station Committee on Organization and Policy of the Association of Public and Land-Grant Universities and its corollary in Cooperative Extension, the Extension Committee on Organization and Policy, formed the NIPMCC. The NIPMCC brings together programs funded under the Crop Protection and Pest Management program at NIFA. Projects are funded competitively at 1862 and 1890 institutions, and can have **state or regional scope**, and focus on managing plant pests.

The existence of NIPMCC **facilitated awareness, communication, and collaboration** among disparate parties. It shows the resourcefulness of the scientific community in taking steps to organize itself across the public and private sectors. Another broader effort, called the Tactical Sciences Coordination

[4] See https://foundationfar.org/grants-funding/grants, accessed September 20, 2022.

Network,[5] was launched to inspire collaboration and is also meant to incorporate a broader suite of NIFA-funded **biosecurity programs in plants, animals, and disaster response**. It was started at the beginning of the COVID-19 pandemic and is gaining momentum. Although most of these NIFA-sponsored programs are hosted at 1862s, some have partners at 1890s. The main intent, however, is to build collaboration among programs regardless of the host institution type. Apart from the NIPMCC and Tactical Sciences Coordination Network, there is a large number of coordinating committees trying to build national cohesion around a variety of issues of national importance.

Management of Zebra Chip to Enhance Profitability and Sustainability of U.S. Potato Production

Another respondent to the Panel's preliminary observations mentioned the Specialty Crop Research Initiative (SCRI) as an important competitive grants program administered by NIFA to bring science to bear on the urgent needs of producers through coordination and collaboration. As an example, the *Management of Zebra Chip to Enhance Profitability and Sustainability of U.S. Potato Production* centers a coordinated group of researchers on a focused task. It was directed at a newly arrived (at the time) but little known potato disease. The disease is caused by a previously undescribed bacterium and transmitted to potato plants by an insect known as the potato psyllid. It is important because it has caused **millions of dollars in losses** to the potato industry (Munyaneza, 2012). In its early stage (2009–2014), the project investigated methods for control of the potato psyllid, which was **the issue of highest interest to potato producers**. Initially funded through the SCRI at NIFA, researchers determined that because even large psyllid populations may not carry the disease bacterium, pesticide spraying is not always needed to fight the disease, **advancing knowledge** with direct implications for **reducing the environmental impacts** of pesticides and producers' costs.

The project developed experimental treatments that enabled spraying at the most appropriate times for psyllid control using the best sequence of pesticides to optimize effectiveness. The treatments developed also provided **economic benefits** by reducing costs by about $300 per acre.

The project was multidisciplinary, encompassing plant pathology, entomology, horticulture, and economics, and included both **research and extension**. Its methods were both field and laboratory based, and employed molecular techniques as well as traditional field experimentation. The project team included 20 multidisciplinary researchers and extension specialists from six states (California, Idaho, Nebraska, North Dakota, Texas, and Washington) as well as an advisory board of stakeholders and team members from the USDA's ARS. This project ended in 2014, at which time a great deal more work was needed on combatting zebra chip. Further research at Texas A&M that has been supported by NIFA and the Foundation for Food & Agriculture Research continues work on developing long-term solutions to control zebra chip (Mora et al., 2021). The recommendations from the project were **communicated to stakeholder groups**, and the broad geographical representation of the team assured that the messaging would be shared widely. Post-grant, the zebra chip research team continues with second-generation researchers and many of the original team remaining involved through multistate research projects on potato diseases. However, there are no 1890 or 1994 institutions involved in this work.

HISTORICALLY BLACK COLLEGE AND UNIVERSITY CENTERS OF EXCELLENCE

Panel member Moses Kairo introduced the Panel to the HBCU Centers of Excellence (COEs). As part of the 125th Anniversary of the Second Morrill Act in 2015, which celebrated a common mission and purpose among the 19 1890 institutions, three integrative COEs were established, although funds ($5 million) were not appropriated through the Farm Bill until 2018. The COEs were established to leverage dis-

[5] See https://tacticalsciences.org, accessed September 20, 2022.

ciplinary diversity among institutions to deliver positive impacts across the continuum of research, education, and extension. Initially **the COEs focused on increasing diversity in the fields of science, technology, engineering, agriculture, and mathematics; increasing profitability and jobs in underserved farming communities; and addressing global food and nutritional security challenges.** The 2018 Farm Bill authorized six COEs to include additional focus areas, and in 2021, the U.S. Department of Defense awarded funding to two new HBCUs to establish COEs in biotechnology and materials science. The HBCU COEs embody purposeful and focused collaboration across the 1890 institutions. Their success is critically dependent on the continued availability of resources and on successful efforts to work together while recognizing, and overcoming, the challenges of working across issues, institutions, and geographies.

INTRASTATE COLLABORATION

The Panel found that federal funding that catalyzes and supports collaboration at the state level through the land-grant system affords greater opportunities for 1862, 1890, and 1994 institutions to work together. The Agricultural Research, Extension, and Education Reform Act of 1998 requires states to submit work plans and to integrate research with extension activities as a requirement for receiving federal funds. For example, through the Alabama Agricultural Land Grant Alliance, statewide extension activities are jointly conducted by Auburn University (an 1862) and Alabama A&M University and Tuskegee University (both 1890s).

The Montana State University College of Agriculture (COA), Montana Agricultural Experiment Station (MAES), and Extension cooperatively design and implement programs that best align with Montana's sovereign Indian Nations (H. Michael Harrington, personal communication, September 1, 2022). Because this demographic is largely underserved and underrepresented, programs and goals are targeted to generate strong and **beneficial interactions for Montana reservation struggles, priorities, and needs.** Activities include work with Montana's seven 1994 Tribal Colleges and Universities and with tribal councils and colleges across the Rocky Mountain region. Cooperative efforts **provide resources and training in livestock management, childhood obesity, food preservation and safety, pasture restoration, environmental stewardship, sustainable agricultural practices, resource and risk management, pesticide certification, and more.** Cultural sensitivity and inclusiveness is a priority for all COA, MAES, and Extension programming.

In the cases that the Panel was aware of, activities at the state level appear to be much more oriented toward relationship building as a precursor to greater collaboration involving 1862 and 1994 institutions. A mini-workshop organized by the Panel featured several examples, including two described here.

Panel member Wendy Powers noted that several key principles guide the Virginia Tech (an 1862) and Virginia State University (an 1890) Cooperative Extension team:

- Share a common mission for the Commonwealth of Virginia;
- Recognize that the relationship of the two institutions is not a competition and that there is no win if the other suffers as a result;
- Take time to develop the inter-institutional relationship;
- Function as one organization;
- Maintain constant, open communication; and
- Acknowledge the resource disparity between institutions and build ways to transcend it.

Successes of the program include the fact that 12 research stations welcome all faculty across both institutions, the university presidents talk to each other, and there is recognition that participants must remain mindful of the disparate resource allocations and do not require everything to be "equal."

Independently of the extension program, the Institute for Critical Technology and Applied Science at Virginia Tech initiated the Diversity and Inclusion Seed Investments Program to build other research partnerships between Tech faculty and faculty at HBCUs and minority-serving institutions (Jalali et al., 2021).

Examples of outcomes of these efforts include more courses being co-taught, evaluation metrics being woven into campus policies and operations, co-funding of an extension agent position with potential for another one, and **an unwavering commitment to advocating for extension as a whole before the state general assembly**. These outcomes led to forming a multicultural alliance; fostering honest conversations; strengthening the position of the alliance to advance changes in the state for all NIFA priorities; and **establishing a broader reach to Virginia residents** due to increased expertise, experience, and relationships among team members.

The Michigan Inter-Tribal Land Grant Extension System (MILES), introduced by Panel member Steve Yanni and featured in one of the Panel's mini-workshops, encompasses all four land-grant colleges and universities in Michigan, all 12 federally recognized Tribal Nations, and NIFA. These entities are partnering to coordinate an integrated land-grant system in Michigan and to **promote collaboration to determine how to best serve all Tribal Nations, communities, and their citizens**. The creation of MILES recognizes the need for building trust among the partnering organizations, ensuring that all partners have a place at the table, and signifying a genuine desire to work together. It also provides a foundation to establish flourishing relationships and collaboration as well as a potential pathway for reducing administrative barriers caused by differences in management practices and policies among member institutions. Some of the goals of MILES include **addressing programmatic and research requests; developing leadership; conserving natural resources; conducting family and consumer science; and promoting health, nutrition, and economic development**.

DISCUSSION

The Panel found many examples of successful ongoing collaborations involving land-grant colleges and universities that focus on issues of global, national (including Tribal Nations), regional, and state importance. Many of the projects are multidisciplinary. Some have a targeted focus and others serve a coordinating function. The HBCU COEs, while modestly funded, are focusing on building capacity and a breadth of expertise across the HBCU community. The Panel also found examples of systems-level projects in both the multistate and CAP programs.

The 1890 and 1994 institutions are not substantively involved in the multistate activities. There are a number of state-level collaborations, particularly with respect to extension, that are explicitly focused on fostering mutual support and coordination among institutions of all kinds.

Considering this partial snapshot of what is taking place in the land-grant system, the Panel asked several questions. First, are these projects successful and by what metrics? It was beyond the scope of the Panel's task to conduct detailed evaluations of these projects. However, the Panel observed that many of the projects have demonstrated outcomes, strong administrative structures with oversight, reporting, and support from leadership. Many are also long-standing and have evolved (or matured) into another form after their initial program funding. Collaborators on some of the projects continue to co-publish and work together, suggesting that important interpersonal bonds were established in the course of the activity.

Second, are these projects meeting "the needs of the nation and global food security" as the congressional directive asks? The answer is no, at least not entirely. Projects that do not engage key parts of the land-grant system are, by definition, missing the benefits that broader participation would bring and also are not likely to be serving all of the relevant stakeholders in that work. In addition, not all collaborations are addressing problems at the same scale, nor does it seem logical that they need to. The difficulty in answering this question is that a comprehensive picture of the portfolio of work embodied by these collaborations as juxtaposed against the needs of the nation and for global food security does not exist, much less include the broader universe of activity in which the non-land-grant colleges and universities and other organizations participate.

The last sentence of the Panel's Statement of Task seems to recognize this reality, asking the Panel "to recommend processes that Land-Grant schools can use to capture and share successes, outcomes, and impacts of joint projects." Although all of the projects described in this chapter have some kind of informal (website) or formal reporting mechanism and NIFA is able to pull reports on topics of activity, the ability to bring together all of the outputs and relevance of these collaborations would be incredibly powerful. However, it would require a new paradigm for synthesizing information.

Along those same lines, the Statement of Task asks the Panel to recommend actions to enhance the success and impact of inter-institutional activities. In the next chapter, the Panel presents information about three important considerations for enhancing the success and impact of collaborative activities: cutting-edge science, human and organizational capacity, and financial support.

3

Framing the Value of Collaborative and Multidisciplinary Research

As the Panel examined the study Statement of Task, two phrases—"enhance the success and impact of inter-institutional activities" and "on national needs and global food security"—stood out as a higher-level context for the goals of study. Ronnie Green, a member of the Panel, offered the view that "Congress wants us [the land-grants] to tackle the really big questions."

The Panel recognizes that having greater impact—the power to create change—depends on being able to use cutting-edge capabilities from across the scientific and engineering disciplines, but that assembling the intellectual and organizational capacity to exploit those tools for real-world applications requires teamwork, often at a systems-oriented, cross-sectoral level. Funding is also necessary to support such work. This chapter discusses these three ingredients in the context of collaboration in the land-grant system.

SCIENTIFIC CONVERGENCE

The land-grant system has an opportunity to embrace and capitalize on the advances and advantages made possible by multidisciplinary approaches to problem solving, as described in a 2009 report from the National Research Council called *A New Biology for the 21st Century* (NRC, 2009). The *A New Biology* report was prepared in response to a request to examine the current state of biological research in the United States and recommend "how best to capitalize on recent technological and scientific advances that have allowed biologists to integrate biological research findings, collect and interpret vastly increased amounts of data, and predict the behavior of complex biological systems" (NRC, 2009). The report was among the first to recognize that the "essence" of the "new biology" is the reintegration of the many subdisciplines of biology as well as the integration of biology with physics, chemistry, computer science, engineering, and mathematics. This integration would enable the creation of a research community with the capability to tackle a broad range of scientific and societal problems, including those associated with food, environment, energy, and health. However, the report's authoring committee also pointed out that fundamental questions in all areas of biology, from understanding the brain to carbon cycling in the ocean, would become more tractable as the new biology grows into a reality that can contribute to advances across the life sciences. The report received considerable attention and was part of the foundation for a new vision of biological research.

However, although the idea of integrating biology with other disciplines had great appeal across the various scientific communities, it was not necessarily straightforward to adopt this new approach. In 2014, the *A New Biology* report was followed by another National Research Council report called *Convergence: Facilitating Transdisciplinary Integration of Life Sciences, Physical Sciences, Engineering, and Beyond.* The report defines convergence as

> an approach to problem solving that cuts across disciplinary boundaries and integrates knowledge, tools, and ways of thinking from life and health sciences, physical, mathematical, and computational sciences, engineering disciplines, and beyond to form a comprehensive synthetic framework for tackling scientific and societal challenges that exist at the interfaces of multiple fields. (NRC, 2014, p. 1)

The purpose of this second study was to address the challenges to creating and sustaining environments that foster convergence, to better understand these challenges, and to explore examples of ongoing successful convergence programs to inform investigators and organizations interested in expanding or establishing their own efforts. The chair of the authoring committee, Joseph DeSimone, noted,

> The approach embodied by convergence provides a framework for thinking about the research enterprise and the network of partners that together form the ecosystem that enables science from innovative research to translational application. Convergence provides us with an opportunity not only to discuss strategies to advance science but also to elevate discussions on how to tackle fundamental structural challenges in our research universities, funding systems, policies, and partnerships. (NRC, 2014, p. viii)

APPLICATIONS IN AGRICULTURE

How do these ideas and developments relate to the land-grant system? The nature of key questions in food and agricultural science is evolving, and the scientific approaches to address them are increasingly at the convergence of multiple disciplines, use information collected dynamically across multiple scales or geographies, and require advanced data science capability. Research, teaching, and extension that use a systems perspective supported with data science expertise and capability are needed to address the multifaceted problems now facing the agricultural and food systems. For example, while traditional research and extension may have focused primarily on improving crop yields, current questions are more broadly framed, for example, on how to improve crop yields and nutritional benefits in a changing climate and/or without environmental degradation.

Science Breakthroughs to Advance Food and Agricultural Research by 2030 (NASEM, 2019) notes that agriculture faces serious problems today that are unlike those in the past. The natural systems on which agriculture depends are showing serious signs of stress, including water scarcity, increasingly variable weather, floods, and droughts brought on by climate change, land use change, population growth impacts, and other factors. Productivity growth for staple crops is stagnating or declining worldwide, a "warning sign" that new methods will be required to address the continuing need for increased productivity. Neither is the U.S. food supply immune to the impacts of pandemics, conflicts and wars, foodborne illness and the threat of pests and pathogens to crops, livestock, and poultry. U.S. farmers and producers need more tools to manage the pressures they face.

The *Science Breakthroughs* report defined the major goals for food and agricultural research in the next decade to include "(1) improving the efficiency of food and agricultural systems, (2) increasing the sustainability of agriculture, and (3) increasing the resiliency of agricultural systems to adapt to rapid changes and extreme conditions" (NASEM, 2019, p. 2).

Underpinning these goals are major research challenges. How can we improve nutrient use efficiency in crop production systems? How will we reduce soil loss and degradation, mobilize genetic diversity for crop improvement, optimize water use in agriculture, improve food animal genetics, develop precision livestock production systems, improve nutritional value of agricultural products, detect and prevent plant and animal diseases and foodborne pathogens, and reduce food loss and waste throughout the supply chain?

The *Science Breakthroughs* report suggests that some answers lie in finding ways to employ recent scientific and technical advances that offer new opportunities to carry out systems-level approaches in food and agriculture research. Chapter 9 of the *Science Breakthroughs* report, Strategy for 2030, describes how developments in computing, information science, machine learning, materials science and electronics, genomics and gene editing, and behavioral and cognitive science have broad application in agriculture. (Box 3-1 lists these crosscutting "breakthrough" tools.) For example, advances in molecular biology now allow for the development of new food sources and traits to increase resistance to insect pests and diseases; increase nutritional components; increase yields, and nutrient and water use efficiencies; and increase the ability to withstand weather extremes.

Microelectronics and nanotechnology provide improved capabilities to sense and monitor physical, chemical, or biological properties and processes to improve food production sustainability; enable and automate controlled environments for production; control microbes to improve food safety and minimize food waste; and create new materials to monitor and improve animal health. Breakthroughs in robotics and drones do not just enable research but provide producers with tools with which to assess and address field conditions and apply pesticides more precisely, thus reducing environmental impacts.

BOX 3-1
Breakthrough Tools for Systems-Level Research

To be effective, the scientific community should pursue convergent and interdisciplinary research approaches using novel tools to "break through" the difficult problems facing agriculture. These approaches and tools include the following:

1. Use of a systems approach to understand the nature of interactions among the different elements of the food and agricultural systems to produce increased overall efficiency, resilience, and sustainability.
2. The development and validation of highly sensitive, field-deployable sensors and biosensors to enable rapid detection and monitoring capabilities across various food and agricultural disciplines.
3. The application and integration of data sciences, software tools, and systems models to enable advanced analytics for managing the food and agricultural system.
4. The ability to carry out routine gene editing of agriculturally important organisms to allow precise and rapid improvement of traits important for productivity, quality, and climate resilience.
5. Understanding of the relevance of the microbiome to agriculture and harnessing this knowledge to improve crop production, transform feed efficiency, and increase resilience to stress and disease.

Systems approaches to addressing complex problems require the convergence of disciplines across the biological sciences, engineering and technology, the human sciences, and economics. Basic, applied, translational, and data sciences (see Box 3-2) have roles to play in developing systems-based solutions for agriculture.

Artificial Intelligence Research Institutes

Data science, including machine learning and artificial intelligence (AI), is an essential capability for successfully addressing the most important opportunities and challenges in food and agriculture. The *Science Breakthroughs* report explicitly noted the pivotal role of advanced data analytics. In 2020, the National Institute of Food and Agriculture (NIFA) and the National Science Foundation contributed funds to establish seven new collaborative AI institutes to accelerate research, expand America's workforce, and transform society in the future in terms of food and agriculture.[1] In a mini-workshop (see Appendix B) organized by the Panel, participants from two of the institutes explained their work.

One is the Artificial Intelligence for Future Agricultural Resilience, Management, and Sustainability Institute, led by the University of Illinois at Urbana-Champaign with partners at Tuskegee University, Michigan State University, the Donald Danforth Plant Science Center, the U.S. Department of Agriculture's (USDA's) Agricultural Research Service, and the Argonne National Laboratory. The partners are using shared AI tools to explore projects that range from the use of robotics for small-scale producers to "farm of the future" modeling for resilience and sustainability. The second, the AI Institute for Food Systems, based at the University of California (UC), Davis, with partners from UC Berkeley, Cornell University, and the University of Illinois, is pursuing an ambitious array of AI-supported projects around agricultural production, nutrition, food processing, molecular breeding, and importantly, the development of AI specific to the challenges of food and agriculture. Both projects have strong teaching, research, and extension elements.

[1] See https://www.nifa.usda.gov/about-nifa/press-releases/usda-nifa-nsf-invest-220m-artificial-intelligence-research-institutes, accessed September 20, 2022.

The keynote presenter at the workshop, Tom Andriola of UC Irvine, focused on the flexibility and power of these collaborative platforms. In addition to creating an opportunity to provide computing and data science tools for use by all participants, the investment in these AI institutes can serve as a catalyst to design and implement a scaling strategy and plan for data science that is inclusive, and maximizes the full potential of all land-grant colleges and universities in terms of enhanced collaboration. However, data science will only fulfill its potential if all institutions have an opportunity to contribute and have access to well-curated and integrated data assets on the food and agricultural system (see Box 3-2).

BOX 3-2
The Importance of Data Science

Tom Andriola, vice chancellor for information, technology, and data at the University of California (UC), Irvine, and UC Health provided a keynote presentation on the potential of AI and data analytics to address and solve important practical problems. This was illustrated for the health system by the UC Health Data Warehouse, which was initially created at UC Irvine and was eventually adopted by all 10 of the UC campuses. He illustrated how shared data can be an asset for both research and clinical practice. The availability of warehouse health data from the UC health clinics allowed the development of a predictive model for use with COVID-19 patients and helped determine the most appropriate care for individual patients. As a result, UC Irvine achieved the second highest survival rate for COVID-19 patients in the nation.

Given that there are now enough data and advanced computing capabilities for more expanded purposes in health care, the UC Health Data Warehouse supports the re-emergence of AI, platform evolution, distributed analytics, and "data as a strategy" that can bring multidisciplinary professionals together in "ecosystem" (network) thinking. It helped produce the concept of the "collaboratory," in which data can serve multiple purposes at the same time if people who understand how to put data to work are available. The collaboratory focuses on enablement, connectedness, inclusiveness, and lack of boundaries. It provides a data operating system that serves as a mechanism to facilitate interdisciplinary teams, domain experts, and multi-organizational collaborators coming together to create positive impacts.

Furthermore, they are customizable for different domains, and the model could be adapted to the agricultural and food system. Andriola added the statement that data are a strategy, not just an asset. He also emphasized that data can be used to create knowledge but that data assets require curation, aggregation, and linkage. The knowledge that emerges from data analytics can, and should, be put into practice and these practices exported to facilitate the largest possible data ecosystem. Finally, he stated that it is not the data per se but rather the network that is important. His view was that there is considerable commonality between health care and agriculture, and the model he discussed in his presentation would fit very well with the latter.

CREATING SUCCESSFUL COLLABORATIVE TEAMS

The land-grant system, with its nationwide remit, seems an ideal place to activate the recommendations in the *Science Breakthroughs* report. Many of the approaches focused on systems-level discovery will require regional or national cooperation and planning in addition to multidisciplinary efforts. If successful, systems approaches will produce essential information that will be the basis of new knowledge to inform decision making at various scales, along with new tools to implement those decisions. The broader impacts of collaboration are realized at the point when science catalyzes real change. However, getting to that point involves the complex endeavor of many people working together toward a common goal in a sustained way.

One of the respondents to the Panel's preliminary observations remarked on a research group at Colorado State University where investigators study interdisciplinary scientific teams, in real time, to advance a predictive theory of what makes teams successful. Subsequently, the Panel asked Jennifer Cross, a leader in this field, to give a presentation on the "Science of Team Science." Cross explained that she and her colleagues shadow a team and evaluate the connections between its members over time, creating graphic maps illustrating the frequency and quality of contacts between team members. Periodic surveys of participants capture the nature of relationships by asking participants such questions as which colleague they met with to discuss the project, who they sought out for advice, who they had a beer with, and who they trust. The maps show who connects most frequently.

Cross said that the science of team science shows that individuals destined to lead projects need to have—or develop—a collaborative mindset, openness to learning from others, the capacity to consider divergent perspectives, and interpersonal relationship and communication skills. These values, mindset, and interpersonal skills are the foundation needed for the more complex tasks of team science, particularly as it grows to engage partners in other sectors (Cross et al., 2022; Hall et al., 2018, 2019).

In addition to the capabilities and characteristics of individuals, the success of a team depends on building the capabilities of the team. Team science competencies include several domains: trusted and genuine relationships, communication, collaborative knowledge generation, collective problem solving, team management, and team co-leadership. Doing participatory team science requires commitment and patience because it takes time to build relationships and trust, create a shared vision, and overcome the many barriers to participatory research. According to Cross, to make progress in a collaboration, there is no shortcut around building trust and a shared vision, and she has the data to prove it. Cross and her colleagues monitor the growth and dynamics of teams as new people join and leave and assess the coherence of the members of the collaborations by using an individual's stated satisfaction with the project, including the feeling that he/she is contributing and the sense that members share a common vision.

Food systems research includes a diverse array of scientific fields—soil and crop science, animal and meat science, economics, public health, nutrition, engineering, ecology, and rural sociology—and stakeholders who will require the team to fully understand and appreciate the context of their situation. This diversity in scientists and stakeholders requires that some team members cultivate the ability to be "boundary crossers" who help translate science to community members and scientists from other fields as well as understand the diverse perspectives of others. That may not be the most scientifically productive individual but the one on which the success of the collaboration depends. One of the primary tensions to be resolved is that between scientists seeking the gold standard of scientific research versus team members who see the value of feasible research that can meet stakeholder expectations and timelines. Researchers who strive to use perfect data and frontier scientific methods may not be a good fit at the center of a team and may be best engaged as peripheral members.

Cross found that large transdisciplinary teams benefit from participating in evaluation and training activities designed to enhance their capacity to address challenges and foster self-awareness about best practices for working in teams. "Developmental evaluation" is an approach to team development and assessment focused on continuous improvement by supporting team adaptation under dynamic and evolving conditions in real time. Her work is popular at Colorado State University, where science teams actively seek resources to support her work with them.

The Panel is aware that participants face barriers that affect their ability to fully participate in collaborations, and Cross and her colleagues suggest that the results of those stresses can be captured by their research method. This scientific approach to building self-aware multidisciplinary teams is helpful in illuminating the way forward.

CAPACITY FUNDING OF LAND-GRANT COLLEGES AND UNIVERSITIES

Funding is, and will be, essential to conducting inter-institutional collaboration in the land-grant system. The federal government provides money to the land-grant colleges and universities in the states according to a complex set of legislative authorities and formulas to support the system's tripartite mission of agricultural research, education, and extension. NIFA distributes funds to the states as capacity grants, sometimes called formula funds, in amounts based on the proportion of rural and farm population in the state or, relative to the 1994s, based on numbers of Native American students.

Separate from the capacity grants, NIFA also awards funds on a competitive basis through the Agriculture and Food Research Initiative (AFRI) to the highest-quality proposals submitted by a large pool of eligible institutions and organizations that include many non-land-grant entities. AFRI grants cover foundational and applied projects, including small and large collaborative projects, such as the Coordinated

Agricultural Projects. Congress has approved funding of $445 million for AFRI for fiscal year (FY) 2022 (CRS, 2022a). Success rates for obtaining an AFRI grant vary with each specific program and range from 10 to 40 percent (NIFA, personal communication, September 15, 2022).

The legislative authorities that provide funds to the land-grant system are specific to each category of land-grant institution. The Hatch Act of 1887 and the Smith-Lever Act of 1914 fund research and extension, respectively, at the 1862 institutions.[2] In addition to other provisions, these funds must be matched 1:1 by the states with nonfederal funds, and 25 percent of the capacity portion of each of the funding lines must be used for collaboration with one or more states.

The Evans-Allen Act of 1977 and the National Agricultural Research, Extension, and Teaching Policy Act (NARETPA) of 1977 fund research and extension activities, respectively, at the 1890 institutions. Unlike the Hatch Act and Smith-Lever Act funds, there is no required set-aside for collaboration. However, these funds must also be matched 1:1 with nonfederal funds. If 100 percent of matching funds is not found, the 1890 institutions can apply to NIFA for a waiver of up to 50 percent of the matching requirement. Although a waiver allows the university to receive the federal funds, it reduces the overall amount it would have received had the full match been obtained. The 1890s have had to leave money on the table every year since Congress began to require 100 percent matching funds in 2007. In 2020, 10 of the 19 1890 institutions did not meet the 100 percent match; as a result, they collectively "lost" $21 million that otherwise would have been realized (CRS, 2022b; see Table 3-1).

TABLE 3-1 FY 2020 Appropriated Federal Capacity Funding and State-Matching Funds

Program	Federal Funds	State Match	Difference (amount lost)
1862 Hatch Research	$183,367,802	$173,987,802	$9,380,000
Hatch Multistate	$60,020,215	$59,147,425	$872,790
Smith-Lever 1862 Extension	$304,649,217	$293,825,447	$10,823,770
Evans-Allen 1890 Research	$62,910,320	$50,572,520	$12,337,800
1890 NARETPA Extension	$54,720,000	$45,539,245	$9,180,755
Tribal College Research Grants	$4,000,000	NA	NA
Tribal Colleges Extension	$8,500,000	NA	NA
Trial College Endowment Interest (discretionary)	$5,000,000	NA	NA

SOURCE: CRS, 2022b.

The fact that, in the same state, 1862 institutions with primarily White students have been able to obtain matching funds from state legislatures while 1890 institutions with primarily Black students have not is a striking and disturbing inequity. Tennessee withheld nearly $544 million in matching payments over several decades to its sole 1890 institution, Tennessee State University, a situation the legislature is now attempting to redress. In 2021, Maryland settled a lawsuit for $577 million with the state's four Historically Black Colleges and Universities (HBCUs), one of which is an 1890 institution, claiming the state "persistently underfunded" the HBCUs while supporting programs at other institutions, which had the effect of pulling students away from the HBCUs (NEA, 2022).

The consequences of this inequity are apparent in many ways. For example, in a 2018 study, the U.S. Government Accountability Office (GAO) examined capital project needs at HBCUs. The HBCUs responding to GAO's survey reported that 46 percent of their building space, on average, needs repair or replacement; GAO identified capital project needs at nine HBCUs that it visited in the areas of deferred maintenance, facilities modernization, and preservation of historic buildings (GAO, 2018). The U.S. Department

[2] A portion of Hatch funds supports the State Agricultural Experiment Stations associated with the 1862 institutions. Smith-Lever funds support the partnership of the 1862s in Cooperative Extension, a partnership of the land-grant system with federal, state, and local agencies to bring knowledge to practitioners.

of Education's HBCU Capital Financing Program provides access to needed funding designed to help address capital project needs for some HBCUs and has helped modernize their facilities to improve student recruitment, but fewer than half of HBCUs have used the program, according to U.S. Department of Education data.

The situation is even worse for the 1994 institutions. Slightly more than 70 percent of Tribal College and University (TCU) funding comes from the federal government. Because of the government-to-government relationship between the tribes and the federal government, states have no obligation to provide funding to the TCUs and most do not (Nelson and Frye, 2016). In addition, under the Tribally Controlled Colleges and Universities Assistance Act of 1978, the federal government can authorize funding streams for the TCUs, but what has actually been appropriated has been only about half of authorized amounts.

The Tribal College Endowment Program provides 1994 institutions with discretionary funds to support agriculture and the mechanical arts with interest on the 1994 Institutions Endowment Fund, distributed annually according to a formula related to the number of Native American students. The Tribal College Research Grant Program provides grants for research in cooperation with specified types of institutions (including 1862 and 1890 institutions). The Tribal College Extension Grants Program supports 1994 institution extension activities.

Another perspective on funding across the land-grant system is to view it relative to the number of institutions in each category and the number of students they serve. Table 3-2 provides this information, showing that on a per institution basis, the 1862 institutions receive more capacity funds for research than the 1890 institutions and vastly more than the 1994 institutions. However, they also collectively serve many more students.

It is also worth noting that because the allocation of Hatch and Evans-Allen capacity funds are based on the farming and rural population of states, smaller 1862 institutions, such as those in the northeastern United States where the number of farms is lower, also find funding levels insufficient to support their research and extension activities (per comment on the Panel's preliminary observations).

THE LARGER FUNDING PICTURE

Inequities notwithstanding, there is a larger backdrop to the picture of resources for the work of the land-grant system and its potential to engage in greater collaboration, and that is the fact that public-sector (federal and state) funding for agricultural research (adjusted for inflation) has been in decline for two decades. According to researchers at USDA's Economic Research Service, expenditures peaked in 2002 at \$7.64 billion (in 2019 dollars) and by 2019 were \$5.15 billion, about the same level as in 1970 (Nelson and Fuglie, 2022). This downward trend puts pressure on the entire land-grant system, in spite of private-sector investments in agriculture and food of more than double that of the public. It has been acknowledged that private-sector funds are not a substitute for public support as they focus on profit generating activities, whereas public-sector funds address public goods that arise from food and agriculture, including economic returns (CRS, 2022a). Numerous economists have estimated that public funding of agricultural research provides a minimum return of 20:1 for each dollar invested (Nelson and Fuglie, 2022).

TABLE 3-2 FY 2022 Selected Federal Research Funding by Institution Type

	Institution Type		
	Hatch Act	Evans-Allen Act	Tribal Colleges Research Grant Program
Total Appropriation	\$260.0 million	\$80.0 million	\$4.5 million
Total Number of Institutions	57	19	35
Average Appropriation per Institution	\$4.6 million	\$4.2 million	\$0.1 million
Total Number of Students	1,853,496	92,004	15,485

SOURCE: CRS, 2022a, Table 3.

4

Principles for Enhancing the Success and Impact of Land-Grant Collaboration

The Statement of Task asked the Panel to provide a framework of principles for successful coordinated and collaborative activities; to recommend actions to reduce barriers to effective and impactful collaboration; and to recommend processes that the land-grant colleges and universities can use to capture and share successes, outcomes, and impacts of joint projects.

As noted in earlier chapters, individual projects have their own unique characteristics, community, history, scope, and goals. A common vision is likely to play a determining role in the success of any team whether a small collaboration or a large one, but the stage for large systems-level collaborations has to be set at a higher level. Based on its members' own experiences, stakeholder input on its preliminary observations, and insights from the mini-workshops, the Panel offers the following principles with respect to large systems-level projects that could also apply to collaboration more generally:

Support and advocacy for large-scale collaborative projects has to come from the highest administrative levels in the various institutions. Institutions and their leaders need to believe in the importance and value proposition of collaborative activities and articulate that support to their internal and external communities. Leaders ought to be willing to talk to external stakeholders about the reason for the project, participate in establishing trust among institutions, develop a common vision of the aspirations and perceived benefits of the activity, and find ways to lower the barrier for faculty participation by negotiating agreements to reduce administrative practices. Such agreements would also send a signal to faculty that there is buy-in from leadership to support collaboration.

Planning is essential. Financial resources are necessary to enable planning among institutions. Planning is a methodical process for envisioning collaborative activities, finding partners, and understanding what is needed for successful implementation. Planning may also require evaluating the assets and opportunities available for the project, such as special infrastructure or access to common platforms for data sharing and analysis or administrative reporting. Finally, planning should include anticipation of not only research outcomes but also considerations to maximize impact through education and extension and outreach. For example, planning data management at the outset of a project could expand its impact through effective use of data repositories, open models, or other platforms that "democratize" data for the broader research community. To assist in planning, the National Institute of Food and Agriculture (NIFA) could fund preliminary studies or staged preproposals at modest cost that require input from diverse institutions and organizations.

Focused operating resources should be adequate to support the participants of the project, including administrative support and trained project and data management support, in addition to directed funds to support research, teaching, and outreach. The integrated elements of research, teaching, and outreach should have dedicated support in land-grant activities, and the plan for these needs to be developed in the budget for the project. Large-scale projects should consider teaching and extension to be central to the purpose of the effort and to the actual impact of the project. A well-supported project will be much more likely to succeed and have impact.

Incentive structures are needed to enable faculty to participate in large-scale projects without compromising on their other responsibilities. Faculty member time, especially at smaller institutions, may already be fully committed to teaching and other assignments, leaving insufficient time and resources to take part in collaborative research. In the 1890s and 1994s, and even in some 1862s, fewer faculty members are generally available to share teaching, research, and extension responsibilities. A stakeholder noted that most faculty at Tribal Colleges and Universities (TCUs) are already teaching too many classes and are also overcommitted with administrative duties, which severely limits their ability to take part in research projects. Creative solutions are needed to help find funding for those who have heavy teaching responsibilities to "buy themselves out" while involved in collaborative research, and providing "teaching postdocs" to free up faculty time needs exploration. Ensuring adequate start-up resources for newly hired faculty and investing in "continuation" resources for newly promoted faculty may also offer workable solutions.

Large-scale projects need support for uniform, shared data management systems that enable seamless access to information. The incompatibility of computer systems between institutions leads to barriers in the flow of information. Because it is information that is of the utmost value in any collaboration today, shared systems are an imperative.

Outstanding communications support is needed to inform the public, including legislators, of the outcomes of their financial investments. Outreach related to projects, associated with targeted goals for audiences to reach, is frequently worth the investment. The resources needed for outreach mechanisms are minor in the scope of a budget for a large project but can pay dividends. Professional communicators and stakeholders should be engaged in formulating communication messaging for various audiences and strategy for delivering those messages.

RECOMMENDATIONS FOR REDUCING BARRIERS TO COLLABORATION

1. **It is urgent that Congress take action to facilitate the participation of all land-grant colleges and universities in multistate research and extension projects.**

Currently, inter-institutional collaborations of the multistate research projects do not routinely engage faculty from the full range of institutions across the land-grant system. Collaborations that include the 1890s and 1994s occur much less often than collaborations among the older and better funded 1862s. As noted earlier, different authorities guide the allocation of federal funds to 1862, 1890, and 1994 institutions respectively, with different requirements for state-matching support of federal dollars and different stipulations for the use of funds for collaboration. Historical and current funding disparities have prevented many 1890s and 1994s from being full partners in collaborations with the 1862s. If unifying the land-grant system around meeting common pursuits is important, Congress should address the need to provide dedicated funding to support participation in collaborative activities.

Many comments submitted by stakeholders confirmed the Panel's concerns about inequities across the full range of land-grant colleges and universities and provided added insights into how funding inequities among these institutions are perceived. One commenter noted, "The distribution of leadership and funds is not equitable. It can seem as though the 1862 institutions are merely 'checking boxes' to include a minority-serving institution, rather than functioning out of true partnership or collaboration." Specific concerns were raised about the lack of inclusion of the 1994 institutions in collaborative efforts. For example, one commenter said that many "TCU/1994 faculty are mutually excluded from activities/opportunities afforded to 1862/1890 faculty. When it comes to competitive funding, 1994 LGUs [land-grant universities] are often looked upon to be a minority card to be utilized for funding opportunities—and are often not treated as equals."

Another stated that even 1862 institutions in some states are often left out of multistate proposals due to "lack of crop area" or some other geographic factor. For field crops, this results in the "rich getting richer" in the Midwest, while "highly productive and important areas such as the Pacific Northwest and East Coast are left without funding, and their growers are left without relevant data."

2. **Land-grant administrators should examine how to reduce the variability in committed support for faculty participation in collaborative activities.**

Institutions have differing approaches for supporting faculty involvement in collaborative research, which creates varying expectations about the nature of the activity and the role of participants. For example, some institutions use multistate research funds to support salaries while others may use them to support travel to participate in meetings. The Panel recognizes that experiment station and extension directors have reasons for managing their funds as they do, but expectations for the level of participation in a collaborative activity need to be made clearer by parties involved along with decision making around the value of time spent on collaboration.

Different administrative procedures and policies among land-grant colleges and universities for proposals, intellectual property, reporting, and mechanisms for managing funds may create time lags, paperwork burdens, and opportunity costs that discourage collaboration. Differing overhead costs may also reduce incentives to share funds among multiple institutions. Many land-grant colleges and universities also work in the context of a larger institution that is funded differently and has different expectations of faculty. Some respondents to the Panel's preliminary observations felt treated unfairly by their partners leading a collaboration. However, another said, "If each institution is funded individually, the leader has little power to keep the participants working to the common goal. As much as I like the idea of the leader having less to do, eliminating the coordinating power of budget control would lead to weaker collaboration."

The 1890s and 1994s also have less capacity to deal with the administrative burdens associated with collaborations than the 1862s. One suggestion was to consider investing indirect costs to pay for administrative support to improve the success of multi-institutional collaborations. Such support should be available for the grant application, fiscal management, reporting, and other administrative phases associated with procuring and conducting collaborative research projects. The speakers participating in the mini-workshop on the role of capacity for collaboration were supportive of the idea to establish a public or private organization with up-to-date knowledge of business software and procedures to provide administrative assistance to land-grant colleges and universities. It is also worth noting that NIFA allows the inclusion of funding for project management personnel in proposals, a provision of which many large well-funded institutions already take advantage.

3. **Faculty members in academic departments should reflect on how collaborative activities fit into an academic career and advocate for their reward.**

Like many, if not most, academic entities, land-grant colleges and universities have traditions emphasizing and rewarding competitive, rather than collaborative, research projects. This can be a disincentive to collaboration. Pre-tenured faculty, in particular, may encounter strong disincentives for pursuing collaboration. One stakeholder agreed: "We often have people spending time competing for resources rather than collaboratively working toward a goal." Another respondent said, "There is often less understanding of multidisciplinary research and limited understanding of the role of extension in other parts of the university, which impacts promotion and tenure processes." One stakeholder, however, suggested that "the tradition of rewarding competitive rather than collaborative projects has ended." Another said, "Departmental cultures around collaborations vary widely but are constantly changing. In the natural sciences, any institution that has failed to keep up with the collaborative nature of research in their evaluation procedures will fail to be competitive at all." If this is true, then the time has come for departments to modify evaluation criteria for promotion and tenure to ensure that achievements in team science are appropriately recognized and rewarded.

4. **Funding agencies should help faculty find partners for collaboration.**

Several respondents from 1862 institutions to the Panel's preliminary observations acknowledged the problematic nature of not involving participants from the 1890 and 1994 institutions and even expressed dismay at being unaware of the expertise at institutions across the land-grant system, not to mention at non-land-grant institutions or other organizations that may serve as suitable partners for collaboration.

Stakeholders had several suggestions to address this issue. One said,

> I think USDA [the U.S. Department of Agriculture], NSF [National Science Foundation], and other funding agencies can do a better job to teach colleagues how to access previous project

> reports. There is plenty of information in CRIS [Current Research Information System] or other systems to show what was funded, who the team members are, how they work together, and targeted outcomes. Most of the people do not know how to use the reporting systems to find new ideas and to create new collaborations.

Another suggested that "developing programs to allow information exchange in face-to-face venues could be extremely helpful in laying foundations for collaboration. Possible ways to achieve such goals could include facilitating faculty attendance at scientific society meetings and creating expertise registries to assist researchers in identifying potential partners."

To increase knowledge of where there is relevant expertise across the land-grant system, NIFA could encourage collaboration by convening information exchange or sandbox workshops. These could highlight funded research projects or new areas of research and allow investigators to share information about work they may be doing in topical areas that are of interest to other scientists in the system. This could, in turn, help to establish relationships among investigators with common interests in particular problems and, as suggested above, help create interpersonal foundations for future collaborative work.

NSF and USDA's Agriculture and Food Research Initiative fund Research Coordination Networks that support groups of investigators to communicate and coordinate their research across disciplinary, organizational, divisional, and geographic boundaries. These could serve as models for addressing the need for investigator networks. Several individuals from 1994 institutions emphasized the need to build trust and relationships prior to inviting them to be included on a proposal or even to participate in meetings, which makes yet another claim on limited time. The collaborative activities between Tribal Councils and the State Agricultural Experiment Station and Extension in Montana as well as the Michigan Inter-Tribal Land Grant Extension System (see Chapter 2) may be useful models for partnering with 1994 institutions.

5. **Land-grant colleges and universities should help build capacity for collaboration by introducing faculty to the science of team science.**

Participating in collaboration requires emotional intelligence and project management skills, for which many faculty are unprepared and untrained. As noted earlier, having these skills is central to the success of team science (Cross et al., 2022; Hall et al., 2018, 2019). Academic institutions generally only aim to train students in their scientific disciplines, not in understanding the human behaviors that foster successful teams. There is a scientific basis for understanding these dynamics in collaboration, and academic institutions should use that knowledge to inspire more explicit thinking around team dynamics and self-awareness by leaders of and participants in collaborative activities. One respondent to the Panel's preliminary observations pleaded that the Panel not recommend "another useless training exercise." Institutions should pilot ways to expose faculty to the science of team science so that potential for skill-building is supported by credible and actionable feedback. It is a low bar to listen to and learn from the experiences of others.[1]

RECOMMENDATIONS FOR AMPLIFYING IMPACTS THROUGH EFFECTIVE COMMUNICATIONS

1. **NIFA should support a collaborative research and extension activity of social scientists and science communicators in the land-grant system to develop outreach strategies.**

Documenting the economic, environmental, and societal impacts and benefits of collaborative research is a way of raising the profile of this information for producers, policy makers, and the public. Simply stated, impact is the effect that an activity or project, especially something new, has on a situation, person, or policy. Constituents may be more likely to recognize the impacts and benefits of collaborative activities

[1] See https://www.nifa.usda.gov/leading-transdisciplinary-projects, accessed September 20, 2022.

among land-grant colleges and universities if communication strategies are tailored to their interests and levels of understanding. For example, producers are likely to have quite different ideas about what constitutes a successful project than members of the public. Communicating results to producers may require more approaches to emphasize how results address their specific concerns. Policy makers are likely to value the economic benefits of a project while members of the public may consider its impacts on food prices or on environmental effects.

Is there a land-grant "brand" that is used to describe impacts for the land-grant system as a whole? The simple answer is no. Land-grant institution scientists typically report the results of their research in journals or other publications that may not be read by policy makers and the public. Land-grant colleges and universities want and need to be responsive to stakeholders with state and regional priorities but may focus less on distributing information nationwide.

There are means that already exist to increase distribution of research outcome information and expand it nationwide when appropriate. For example, NIFA has a communication unit that collects, writes, and transmits the impacts of the work NIFA supports to the public.[2] The NIFA staff also offers workshops on recognizing and effectively communicating impacts. NIFA's "Share Your Science" campaign[3] is designed to highlight research outcomes and accomplishments on a national level. It is aimed at spotlighting the achievements being made by NIFA's partners in addressing societal challenges, such as increasing food security; decreasing hunger; and addressing climate change, food safety, childhood obesity, and sustainable energy. Nongovernmental organizations, such as the Supporters of Agricultural Research Foundation,[4] also seek to educate stakeholders about the importance of agricultural research and provide information for this purpose.

However, the Panel finds that there is a need for more effective communication strategies targeted to specific audiences to increase impact as well as producer, political, and public support. A new vision for capturing and communicating outcomes from publicly funded research is needed that uses advanced tools and interfaces to generate information for the many different users of that information. Communication about research outcomes should be tailored to the interests and levels of understanding of diverse audiences.

2. **Land-grant colleges and universities should create novel messaging vehicles to reach specific audiences about the outcomes of collaborations.**

Effectiveness can be enhanced by creating specific messages for particular audiences and by using tools that increase audience understanding. For example, the use of graphical tools can be an effective method of communication for some audiences. Telling the story behind the research, perhaps by using analogies, and clearly explaining why the results are important for ordinary citizens or policy makers could be very helpful for nonprofessional audiences. Student-led hackathons could bring a fresh approach to information dissemination. Project impacts could also be broadened by using extension funds to bring knowledge or practices developed in collaborative (and other) projects to wider user audiences and to support the dissemination of information using new formats and tools.

Greater thought on how to measure and communicate the value of research outcomes is needed to increase their visibility. Anticipating expected outcomes and how to communicate them, defining what success looks like, and creating metrics for impacts could increase the effectiveness of communication efforts. The Experiment Station Committee on Organization and Policy (ESCOP) recognizes a Hatch multistate project as an exemplary multistate activity each year. The projects in the competition exhibit high scientific standards and research relevant to regional needs. Award-winning projects must demonstrate high scientific quality, be relevant to a regional agricultural priority, and exhibit multistate and multidisciplinary collaboration and professional leadership in the conduct of the project. The ESCOP awards also emphasize what the winning teams did together that could not have been accomplished only through the work of

[2] See https://nifa.usda.gov/impacts, accessed September 20, 2022.

[3] See https://nifa.usda.gov/shareyour-science, accessed September 20, 2022.

[4] See https://supportagresearch.org, accessed September 20, 2022.

individuals. Publicizing this aspect could help various audiences understand the importance of collaboration. These criteria could be useful in other contexts and could be emphasized in communicating project success and significance. To strengthen the ability to capture and communicate research impacts, NIFA and other funding agencies might also provide added support for post-collaboration assessments. This would help determine which approaches are most effective and should be shared with others who wish to begin new collaborations.

5

Next Steps

Increased collaborative activity across the land-grant system has the potential to expand the reach of its work, and if effectively communicated, increase its relevance to stakeholders, including the private sector, policy makers, and the public. The land-grant system already provides a robust research, education, and outreach infrastructure, supporting a national research portfolio that ranges from short-term projects directed by individuals to longer-term, larger-scale group efforts. Emphasizing collaboration as an additional hallmark of its mission could allow the land-grant universe to be recognized as a more connected national system.

The Panel recognizes that the principles and recommendations in this report provide only an initial framework to enhance collaboration across the land-grant system that would allow it to operate more effectively as a unified whole. Adopting a culture of collaboration implies change in behavior reinforced by enabling policies. Universities and colleges could adopt curricula for project management, communication, and other skills that would improve chances for team success and perhaps provide new and satisfying career pathways for students and young faculty members. If the time required for project planning and team development and maintenance was treated as valuable during tenure and promotion deliberations, faculty might be more likely to pursue collaboration.

Some of the obstacles mentioned in the report are not trivial to overcome. Resources that need to be increased include more than just funding, although that is a crucial one. Others are staffing, available time for faculty to participate in research, and data infrastructure, including data science. Addressing some of the obstacles noted in this report implies an urgent need for additional funding and responsibility in return for greater effectiveness, capacity, and "true" partnership. Financial support for administrative assistance would enhance the ability of teams to develop successful proposals and would help level the playing field, allowing 1890 and 1994 institutions to become genuine participants in collaborative projects.

The various constituencies in the scientific community are encouraged to think broadly about the need to establish more forward-looking mechanisms for nationwide coordination of agricultural and food research. Diverse backgrounds, cultures, community relationships, and scientific expertise reside in the land-grant colleges and universities, but collaborative vehicles are needed to take advantage of that diversity and allow the institutions to operate in a more connected way. Opportunities such as data platforms with artificial intelligence tools, which participants can collectively and simultaneously contribute to and draw from in pursuit of answers to independent questions, are very compelling from the standpoint of knowledge generation around shared resources. There may be other collaborative platforms that could be built as an inclusive resource and a conduit for experimental data.

Collaboration organizers may also not be aware of the potential for external non-land-grant public or private partners with assets, such as data platforms, technologies, funds, and networks, to participate and support land-grant collaborations.

The kinds of partners that can help realize the vision of a project at the start and over time could include individuals from both land-grant and non-land-grant colleges and universities, individual producers and producer groups, state departments of agriculture, foundations and other nongovernmental organizations, international organizations, entrepreneurs and venture capitalists, and others in private industry. While land-grant investigators must be careful to avoid potential conflicts of interest, partnering with industry can have real benefits in helping to foster research that can translate into applications that drive positive economic and social impacts.

Finally, while there are clearly ways to increase the success of collaboration and team science in agricultural research, the success of large projects cannot be achieved without adequate and stable funding for the overall enterprise for food and agriculture research. The Panel hopes that the ideas offered in this report will gain the attention of leaders of the land-grant colleges and universities and their stakeholders, and that they will be willing to spend time developing and implementing them with the support of Congress.

References

Cross, J.E., B. Jablonski, and M. Schipanski. 2022. Inquiry Within, Between, and Beyond Disciplines. In *Food Systems Modelling: Tools for Assessing Sustainability in Food and Agriculture*, edited by C.J. Peters and D.D. Thilmany, pp. 327–348. London: Elsevier Inc. https://doi.org/10.1016/B978-0-12-822112-9.00010-2.

CRS (Congressional Research Service). 2019. *The U.S. Land-Grant University System: An Overview.* Report R45897. Washington, DC: CRS.

CRS. 2022a. *The U.S. Land-Grant University System: Overview and Role in Agricultural Research.* Report R45897, Updated. Washington, DC: CRS.

CRS. 2022b. 1994 Land-Grant Universities: Background and Selected Issues. *In Focus*. https://sgp.fas.org/crs/misc/IF12009.pdf.

GAO (U.S. Government Accountability Office). 2018. *Historically Black Colleges and Universities: Action Needed to Improve Participation in Education's HBCU Capital Financing Program*. GAO-18-455. Washington, DC: GAO.

Hall, K., A. Vogel, G. Huang, K. Serrano, E. Rice, S. Tsakraklides, and S. Fiore. 2018. The Science of Team Science: A Review of the Empirical Evidence and Research Gaps on Collaboration in Science. *American Psychologist* 73:532–548. https://doi.org/10.1037/amp0000319.

Hall, K.L., A.L. Vogel, and R.T. Croyle, eds. 2019. *Strategies for Team Science Success: Handbook of Evidence-Based Principles for Cross-Disciplinary Science and Practical Lessons Learned from Health Researchers*. Cham, Switzerland: Springer. https://doi.org/10.1007/978-3-030-20992-6.

Jalali, Y., C. Matheis, C. Tysor, and V.K. Lohani. 2021. Sustaining Faculty Collaboration: An Exploratory Process-Based Study of Research Collaboration Across Universities. American Society for Engineering Education Paper 28454. Washington, DC. https://monolith.asee.org/public/conferences/170/papers/28454/view.

Kopp, R.E. 2021. Land-Grant Lessons for Anthropocene Universities. *Climatic Change* 165:28. https://doi.org/10.1007/s10584-021-03029-9.

Mora, V., M. Ramasamy, M.B. Damaj, S. Irigoyen, V. Ancona, F. Ibanez, C.A. Avila, and K.K. Mandadi. 2021. Potato Zebra Chip: An Overview of the Disease, Control Strategies, and Prospects. *Frontiers in Microbiology* 12:700663. https://www.frontiersin.org/articles/10.3389/fmicb.2021.700663/full.

Munyaneza, J.E. 2012. Zebra Chip Disease of Potato: Biology, Epidemiology, and Management. *American Journal of Potato Research* 89(5):329–350. https://doi.org/10.1007/s12230-012-9262-3.

NASEM (National Academies of Sciences, Engineering, and Medicine). 2019. *Science Breakthroughs to Advance Food and Agricultural Research by 2030*. Washington, DC: The National Academies Press. https://doi.org/10.17226/25059.

NEA (National Education Association). 2022. A Looming Crisis for HBCUs? An Analysis of Funding Sources for Land Grant Universities. Issue Brief Number 2, March 15. https://www.nea.org/resource-library/land-grant-university-research.

Nelson, C.A., and J.R. Frye. 2016. Tribal College and University Funding: Tribal Sovereignty at the Intersection of Federal, State, and Local Funding. Issue Brief, Center for Policy Research and Strategy, American Council on Education, May. www.acenet.edu/Documents/Tribal-College-and-University-Funding.pdf.

Nelson, K.P., and K. Fuglie. 2022. Investment in U.S. Public Agricultural Research and Development Has Fallen by a Third Over Past Two Decades, Lags Major Trade Competitors. *Amber Waves*, June 6.

NIFA (National Institute of Food and Agriculture). 2022. Agriculture and Food Research Initiative—Sustainable Agricultural Systems. https://www.nifa.usda.gov/grants/funding-opportunities/agriculture-food-research-initiative-sustainable-agricultural-systems.

NRC (National Research Council). 2009. *A New Biology for the 21st Century.* Washington, DC: The National Academies Press. https://doi.org/10.17226/12764.

NRC. 2014. *Convergence: Facilitating Transdisciplinary Integration of Life Sciences, Physical Sciences, Engineering, and Beyond.* Washington, DC: The National Academies Press. https://doi.org/10.17226/18722.

Appendix A

Panel Biographies

Kathryn J. Boor, PhD, is the dean of the Graduate School and the vice provost for graduate education at Cornell University. Previously, she was the Ronald P. Lynch Dean of the College of Agriculture and Life Sciences at Cornell University. Her research focuses on identifying biological factors that affect transmission of bacteria in food systems. Boor is a fellow of the American Academy of Microbiology, the International Academy of Food Science and Technology, the Institute of Food Technologists, the American Association for the Advancement of Science, and the American Dairy Science Association. She was named a 2018 Woman of Distinction by the New York State Senate. She earned an MS in food science from the University of Wisconsin and a PhD in microbiology from the University of California, Davis. She serves on the board of directors for Seneca Foods Corporation, International Flavors and Fragrance, the United States-Israel Binational Agricultural Research and Development Fund, the Foundation for Food & Agriculture Research, the Science Board for the U.S. Food and Drug Administration, and the New York State Southern Tier Regional Economic Development Council. She served on the Institute of Medicine and National Research Council's Committee on the Review of the Use of Scientific Criteria and Performance Standards for Safe Food.

Olga U. Bolden-Tiller, PhD, assumed the deanship of the College of Agriculture, Environment and Nutrition Sciences (CAENS) and the 1890 Research Director, effective January 1, 2022, at Tuskegee University (TU). Bolden-Tiller was a professor and the head of the Department of Agricultural and Environmental Sciences and the assistant dean for development in CAENS. Bolden-Tiller is a graduate of Fort Valley State University. She has a PhD in animal sciences from the University of Missouri–Columbia and has done postgraduate work in academic leadership at the University of Chicago School of Professional Psychology. Prior to her employment at TU in 2006, she served as a visiting scientist at the University of Texas MD Anderson Cancer Center, an adjunct faculty member in the Houston Community College System, and as support faculty and a postdoctoral fellow at the University of Texas MD Anderson Cancer Center. At Tuskegee, she worked herself up through the ranks from assistant professor to full professor and finally head of the department, having received numerous awards, including the TU Outstanding Faculty Performance Awards for Teaching and Outreach as well as TU's Faculty Achievement Award, the highest honor given to faculty at the institution. She is a successful grant writer and has acquired extramural funding for grants ranging from \$52,000 to \$20 million, supporting the training of numerous high school students and more than 50 undergraduate and graduate students, which has resulted in numerous publications. She holds memberships and leadership roles in several professional associations, including the National Society for Minorities in Agriculture, Natural Resources, and Related Sciences, for which she currently serves as the president.

Dina J. Chacón-Reitzel, MBA, has been the executive director of the New Mexico Beef Council since 1991. She was named in 2003 by the New Mexico State University (NMSU) College of Agricultural, Consumer and Environmental Science (ACES) to serve as one of three delegates to the U.S. Department of Agriculture's Council for Agricultural Research, Extension and Teaching (CARET) for NMSU. She has also been on the National CARET Executive Board for 12 years and chair of CARET for 2 years. In her role as a CARET delegate, she advocates for federal funds in Washington, DC, for the college. She was appointed to serve as the chair on NMSU's Board of Regents for 2 years and currently serves as the vice-chair. Chacón-Reitzel also served on the board of the Western Rural Development Center, a collaboration of the Western Land-Grant Universities. Chacón-Reitzel has served on the National Cattlemen's Beef Association Beef Safety Research Committee for more than 20 years. She has been awarded the New Mexico Outstanding Leadership Award and the Distinguished Alumni Award from the NMSU College of ACES and the Industry Partner Award from the New Mexico Department of Agriculture. Chacón-Reitzel has been honored by New Mexico agriculture organizations, including the New Mexico Cattle Grower's Association's King Service Award and the Northern New Mexico Stockman's Association's Vaquero Award. She is a graduate of Leadership New Mexico. Chacón-Reitzel earned her MBA from NMSU.

Martin A. Draper, PhD, serves as the associate dean for research and graduate programs, College of Agriculture, and the director for research, K-State Research and Extension at Kansas State University. He previously served as the head of the Department of Plant Pathology at Kansas State University and the national program leader for Plant Pathology and Integrated Pest Management with the U.S. Department of Agriculture. He also served as an extension plant pathologist at South Dakota State University (SDSU), where he taught undergraduates and led an applied research program to complement his extension educational programming directed at producers and agricultural stakeholders and decision makers; and the director of the Plant Pest Diagnostic Lab and Seed Health Testing Lab at North Dakota State University (NDSU), where he taught introductory plant pathology and provided service to stakeholders. In his positions, he has had exposure to field crops, horticultural crops, and home garden and landscape plants. He has also consulted with the Serbian government on a restructuring of its extension service and has provided teaching and outreach support in Bolivia. Draper is a member of the American Phytopathological Society, the American Evaluation Association, the American Association for the Advancement of Science, Gamma Sigma Delta, Epsilon Sigma Phi, Phi Sigma, and Sigma Xi. He was recognized by the U.S. Secretary of Agriculture, the SD Agri-Business Association, the SD Soybean Research and Promotion Council, and the SDSU Extension Specialists Association. Draper received his MS and PhD in plant pathology at NDSU.

Ronnie D. Green, PhD, is the chancellor of the University of Nebraska–Lincoln, Nebraska's flagship land-grant and Big Ten institution. He served on the animal science faculties of Texas Tech University and Colorado State University, as the national program leader for animal production research for the U.S. Department of Agriculture's Agricultural Research Service (USDA-ARS), and as the executive secretary of the White House's interagency working group on animal genomics within the National Science and Technology Council. Prior to returning to the University of Nebraska–Lincoln, Green served as a global executive for Pfizer Animal Health's animal genomics business. He is a past president of both the American Society of Animal Science (ASAS) and the National Block and Bridle Club and has served in a number of leadership positions for the U.S. Beef Improvement Federation, the National Cattlemen's Beef Association, the National Pork Board, the National Academy of Sciences, the Association of Public and Land-Grant Universities, and the Alpha Gamma Rho national fraternity. He is a fellow of ASAS and the American Association for the Advancement of Science and was honored with the ASAS Morrison Award, the highest international academic recognition for an animal scientist. He received his MS in animal science from Colorado State University and his PhD in animal breeding and genetics jointly at the University of Nebraska–Lincoln and the USDA-ARS. He is a member of a number of boards, including the Big Ten Council of Presidents and Chancellors, Neogen Corporation, and Supporters of Agricultural Research.

Moses T. Kairo, PhD, is a professor, the dean, and the director of land-grant programs in the School of Agricultural and Natural Sciences at the University of Maryland Eastern Shore (UMES). Prior to joining UMES, he was a professor of entomology and the director of the Center for Excellence in Biological Control at Florida A&M University for nearly 7 years. He also served as the associate research director for its 1890 land-grant institution programs. Kairo has previously worked for CABI, an international development organization, where he served as the lead scientist and a regional representative of the organization's activities in the Caribbean and Latin America region. His research interests have focused on biological control, invasive species management, and crop protection. Over the years, he has carried out work in Africa, Europe, the Middle East, Latin America, and the Caribbean. He has served the land-grant community through various bodies or committees of the Board of Agriculture Assembly of the Association of Public and Land-Grant Universities (APLU). He recently completed his term as the chair of the Experiment Station Section of APLU. He received an MS in applied entomology from Imperial College, London, and a PhD in entomology from the University of London.

Jan E. Leach, PhD, is a university distinguished professor and the associate dean for research in the College of Agricultural Sciences at Colorado State University. She is a plant pathologist who studies the molecular basis of plant disease susceptibility and resistance and how these responses are influenced by interactions with the environment. Leach is the current president of the International Society for Plant Pathology. She is a fellow and a past president of the American Phytopathological Society (APS), a fellow of the American Association for the Advancement of Science, and a fellow of the American Academy of Microbiology. In 2019, Leach was awarded the Agropolis Fondation Louis Malassis International Scientific Prize for Agriculture and Food for Distinguished Scientist, and in 2020, she was presented the APS Award of Distinction, an award that has been given only 17 times in the 112 years of APS's history. Leach was elected to the National Academy of Sciences in 2021. She holds a PhD in plant pathology from the University of Wisconsin–Madison and an MS in microbiology from the University of Nebraska–Lincoln. Leach was the chair of the National Academies of Sciences, Engineering, and Medicine's study on California Agricultural Research Priorities: Pierce's Disease and currently serves as a member of the National Academies' Board on Agriculture and Natural Resources.

Karen Plaut, PhD, is the Glenn W. Sample Dean of the College of Agriculture at Purdue University where she administers teaching, research, and extension programs. She is responsible for the Indiana Agricultural Experiment Station, the Cooperative Extension Service, and a number of state regulatory agencies. She was previously the senior associate dean for research and the director of the Agriculture Experiment Station at Purdue University. Earlier in her career, she was on faculty at the University of Vermont, followed by 2 years as the lead scientist for the National Aeronautics and Space Administration's (NASA's) International Space Station Biological Research project. She then became the chair of the Animal Science Department at the University of Vermont, and later at Michigan State University, before joining Purdue University. As a professor of animal sciences at Purdue University, Plaut leads a research program in mammary gland biology. She has served as a panel reviewer and received funding from the U.S. Department of Agriculture, the National Institutes of Health, the National Science Foundation, and NASA. She was elected as a fellow of the American Dairy Science Association in 2020. Plaut earned her MS from The Pennsylvania State University and PhD in animal science from Cornell University. She is a member of the National Academies' Board on Agriculture and Natural Resources.

Wendy Powers, PhD, is the Cashup Davis Family Endowed Dean of the College of Agricultural, Human, and Natural Resource Sciences at Washington State University, a position she has held since August 2022. She was previously the associate vice president of agriculture and natural resources at the University of California, where she also served as the associate director of Cooperative Extension and the associate director of the Agriculture Experiment Station. Powers started her career as an assistant professor/extension specialist in the Animal Science Department at Iowa State University. After 10 years, Powers moved to Michigan State University (MSU) for another 10 years as a professor/extension specialist and the director of environmental stewardship for animal agriculture at the College of Agriculture, Animal Science and Biosystems & Agriculture Engineering. During this time, she spent 4 years as the first director of the MSU Extension Agriculture and Agribusiness Institute. Powers earned a PhD in animal science and an MS in dairy science from the University of Florida.

Ryan F. Quarles, PhD, is the Commissioner of Agriculture in Kentucky. First elected to that position in 2015, he was re-elected in 2019. Under his leadership, the Kentucky Department of Agriculture started several new programs, including initiatives to combat hunger and connect Kentucky farmers to new markets. Quarles's outspoken advocacy for farmers has been recognized on the regional and national level. He has served as the president of the National Association of State Departments of Agriculture, as the president of the Southern Association of State Departments of Agriculture, and as the chair of the Republican Agriculture Commissioners Committee. From 2010 to 2014, Quarles served as a member of the House of Representatives. His office supports land-grant events with Kentucky State University and the University of Kentucky. Quarles graduated with a master's degree in higher education from Harvard University. He received his law degree from the University of Kentucky and his doctorate in higher education from Vanderbilt University.

Harold H. Schmitz, PhD, is the co-founder and a general partner of March Capital US, LLC, and a senior scholar in the Graduate School of Management at the University of California (UC), Davis. Previously, he served two terms as the Art and Carlyse Ciocca Visiting Professor of Entrepreneurship and Innovation in the Graduate School of Management at UC Davis. He worked at Mars, Incorporated for 25 years, where he developed and executed the company's global strategy for advanced research and innovation at the intersection of food, agriculture, health, and sustainability via collaborations across sectors, including the land-grant colleges and universities. He is an advisor to the Novo Nordisk Foundation's BioInnovation Institute in Denmark and to the Artificial Intelligence Institute for Next-Generation Food Systems located at UC Davis, which is co-funded by the U.S. Department of Agriculture (USDA) and the National Science Foundation. He was recognized as Outstanding Alumnus of the Dale Bumpers College of Agriculture, Food, and Life Sciences at the University of Arkansas, and was elected a Life Member in the Council on Foreign Relations. He was a USDA National Needs Research Fellow during 1990–1993 at North Carolina State University, where he trained in the intersection of food, agriculture, and health and received his PhD in food science. He received his MS in food science from the University of Illinois.

Steve Yanni, PhD, is the land-grant director for Bay Mills Community College (BMCC), a tribally controlled college located in Upper Michigan along the shores of Lake Superior. He has been working with the Bay Mills Indian Community in various capacities for 23 years. Prior to working with Bay Mills, he was a faculty member for 18 years at Lake Superior State University and worked for Michigan State University as a county extension director. In his current position he oversees BMCC's efforts as a 1994 land-grant institution. Current areas of focus include continued development of BMCC's Waishkey Bay Farm as a teaching, research, and incubator facility focusing on sustainable agriculture and food production that is consistent with traditional Anishinaabek values; development of academic programs in sustainable agriculture; community development and health promotion efforts; overseeing BMCC's Mukwa Health and Fitness Education Center; and assisting with the evolution of the Michigan Inter-Tribal Land Grant Extension System. Yanni also serves as BMCC's liaison with the Higher Learning Commission, coordinates research efforts by and with BMCC, and represents 1994 institutions on the Association of Public and Land-Grant Universities' Policy Board of Directors. Yanni earned his PhD from Michigan State University.

Appendix B

Webinar and Virtual Workshop Agendas[1]

COMMITTEE ON ENHANCING COORDINATION BETWEEN LAND-GRANT UNIVERSITIES AND COLLEGES

June 22, 2022
Via Zoom

THE SCIENCE OF TEAM SCIENCE

11:00 am	**Welcome and Introductions** Kathryn J. Boor, *Chair*, Blue Ribbon Panel (BRP), Cornell University
11:10 am	**Presentation: Research Findings on Interdisciplinary Team Science** Jennifer Cross, Colorado State University
11:50 am	**Questions from the Committee**
12:30 pm	**Adjourn**

July 27, 2022
Via Zoom

CAN DATA SCIENCE AND ARTIFICIAL INTELLIGENCE (AI) ENABLE NEW COLLABORATIVE PLATFORMS BETWEEN DIVERSE LAND-GRANT INSTITUTIONS AND CREATE MORE IMPACTFUL OUTCOMES?

11:00 am	**Welcome** Kathryn Boor, *Chair*, Blue Ribbon Panel (BRP), Cornell University **Opening Remarks** Olga Bolden-Tiller, Tuskegee University, and BRP Member (recorded)
11:15 am	**Introduction** Harold H. Schmitz, University of California, and BRP Member **Presentation: Building multi-campus data science platforms to enable advanced data analytics capabilities that accelerate innovation and increase productivity** Tom Andriola, University of California, Irvine
11:45 am	**General Q+A** Harold H. Schmitz, Moderator
12:00 pm	**USDA-NIFA/NSF Artificial Intelligence (AI) Institutes** **Introductions** Harold H. Schmitz (Session Moderator)

[1] Video recordings are available at https://www.nationalacademies.org/our-work/enhancing-coordination-between-land-grant-universities-and-colleges#sectionPastEvents, accessed September 20, 2022.

	Brief Remarks: Steven Thomson, USDA National Institute of Food and Agriculture **AI Institute for Next Generation Food Systems (AIFS)** Gabe Youtsey, University of California, Agriculture and Natural Resources Mason Earles, University of California, Davis **Artificial Intelligence for Future Agricultural Resilience, Management, and Sustainability (AIFARMS)** Gregory Bernard, Tuskegee University Vikram S. Adve, University of Illinois at Urbana-Champaign
12:35 pm	**Moderated Discussion, Q&A**
12:55 pm	**Summary** Harold H. Schmitz and Olga Bolden-Tiller, BRP Members
1:00 pm	**Adjourn**

July 28, 2002
Via Zoom

BUILDING AND SUSTAINING A CULTURE OF COLLABORATION ACROSS THE LAND-GRANT SYSTEM

4:00 pm	**Welcome** Kathryn Boor, *Chair*, Blue Ribbon Panel (BRP), Cornell University
4:05 pm	**Opening Remarks, The Rationale for Collaboration: Four Approaches** Wendy Powers, BRP Member, University of California, Agriculture and Natural Resources
4:15 pm	**Michigan Inter-Tribal Land Grant Extension System (MILES)** Steve Yanni, BRP Member, Bay Mills Community College Emily Proctor, Michigan State University
4:25 pm	**1890s Centers of Excellence** Alton Thompson, Association of 1890s Research Directors Moses Kairo, BRP Member, University of Maryland Eastern Shore
4:35 pm	**Virginia Tech-Virginia State University Cooperative Extension** E.J. Jones, Virginia Cooperative Extension Janine Parker Woods, Virginia State University
4:45 pm	**Multistate Research Project S-1074 Future Challenges in Food Animal Production: Seeking Solutions Through Focused Facilitation** John Classen, North Carolina State University Rebecca Larson, University of Wisconsin–Madison
4:55 pm	**Discussion Session:** Wendy Powers, Moderator • How can collaborations address differences in capacity among the partner institutions and help faculty from across the land-grant system participate fully? • What recommendations do you have for (pick one): institutional leaders, faculty, funding agencies, or Congress to strengthen the cohesiveness and impact of collaborations?
5:25 pm	**Final Remarks** BPR Members Wendy Powers, Moses Kairo, Steve Yanni
5:30 pm	**Adjourn**

August 11, 2022
Via Zoom

THE ROLE OF CAPACITY FOR COLLABORATION IN THE LAND-GRANT SYSTEM

3:00 pm	**Welcome and Opening Remarks:** Kathryn Boor, *Chair*, Blue Ribbon Panel (BRP), Cornell University
3:15 pm	**NASEM Study on Defense Research Capacity at Historically Black Colleges and Universities and Other Minority Institutions** Aliecia McClain, Co-Chair NASEM study committee, Norfolk State University
3:30 pm	**New Mexico State University** Rolando Flores Galarza, New Mexico State University
3:45 pm	**Diné College** Benita Litson, Diné College, Navajo Nation
4:00 pm	**Discussion:** Is collaboration essential for a land-grant college or university? Is it worth the effort? What recommendations do you have for (pick one): institutional leaders, faculty, funding agencies, or Congress to strengthen the attractiveness of collaboration?
4:25 pm	**Final Remarks** Kathryn Boor, BRP *Chair*
4:30 pm	**Adjourn**

Appendix C

Preliminary Observations of the Blue Ribbon Panel

INTRODUCTION

Acting on a congressional directive, the U.S. Department of Agriculture's National Institute of Food and Agriculture (USDA/NIFA) asked the National Academies of Sciences, Engineering, and Medicine to establish a Blue Ribbon Panel to consider how to maximize coordination among universities and colleges of the land-grant system to improve and sustain food security in the United States and beyond and suggest ways to increase and communicate their collective impacts. To that end, the panel established by the National Academies is focusing on the potential to enhance inter-institutional collaboration of participants from a diversity of land-grant institutions—including the 1890s and the 1994s—to increase knowledge generation, solve problems, and create positive opportunities across the food and agricultural knowledge system. In addition, increasing collaboration would also improve both the perception and the reality of the land-grant system as an integrated whole with goals and purposes.

This document contains the Panel's preliminary observations about the nature of collaborative activity across the land grant system and the potential to enhance its impacts. Its purpose is to solicit reactions, ideas, and relevant information from stakeholders who participate in and/or are invested in the outcomes of those activities. Comments on the document can be submitted at https://nap.nationalacademies.org/land-grant-collaboration/form. The comments will inform a workshop being planned for Spring 2022, after which the panel will prepare a brief report to Congress and to USDA with findings and recommendations.

SECTION 1: COLLABORATION IN THE LAND-GRANT SYSTEM

Preliminary Observation #1: There is a significant amount of active and successful inter-institutional collaboration and cooperation taking place in the land-grant system today. Examples of large-scale collaboration among land-grant institutions include the USDA/NIFA-funded Coordinated Agriculture Projects (CAP)[1] and activities funded by the Foundation for Food & Agriculture Research.[2] Other examples of collaborative work come from the Multistate Research Program supported by the federal Multistate Research Fund. Since the mid-1940s, the 1862 Agricultural Experiment Stations have worked together on pressing problems that concern two or more states. Many of these projects have participation from across the system, and include stakeholders from industry, government, and beyond, and in some institutions, faculty are required to be involved in multistate research. The Agricultural Research, Extension, and Education Reform Act of 1998 requires integration of research and extension activities in these collaborations. In addition, there are outstanding collaborations within states as well. For example, in Montana and Michigan, 1862 and 1994 institutions (Tribal Colleges) are working to enhance student and research success as well as outreach. In many states that house both 1862 and 1890 institutions, joint Plans of Work and reports are required. For example, through the Alabama Agricultural Land Grant Alliance (AALGA), statewide extension activities are carried out jointly by Auburn (1862), Alabama A&M, and Tuskegee (1890). These examples of interactions among 1862, 1890 and 1994 institutions are exemplars of collaborative intent, serving as emerging models for enhanced collaboration across land-grant institutions.

---

[1] See https://nifa.usda.gov/afri-regional-bioenergy-system-coordinated-agricultural-projects, accessed October 5, 2022.

[2] See https://foundationfar.org/grants-funding/grants, accessed October 5, 2022.

Preliminary Observation #2: Currently, inter-institutional collaborations do not routinely engage faculty from the full range of institutions across the land-grant system. For a number of reasons, collaboration among the 1862, 1890, and 1994 institutions collectively occurs much less frequently than collaborations among schools within the 1862 group. The 1890 institutions seldom serve as lead institutions for collaborations, while the 1994 schools have very limited resources for this purpose.

SECTION 2: THE RATIONALE FOR COLLABORATION

Preliminary Observation #3: The nature of key questions for food and agricultural science are evolving, and the scientific approaches to address them are increasingly at the convergence of multiple disciplines and involve data collection across multiple geographies. Research, teaching, and extension that use a systems perspective are necessary to address the multifaceted problems now facing the agriculture and food system. For example, while traditional research and extension may have focused primarily on improving crop yields, current questions are more broadly framed; for example, on how to improve crop yields in a changing climate, and/or without environmental degradation.

Preliminary Observation #4: All of us are smarter than one of us: Diversity promotes novel ideas. Given the scope of the problems and challenges that face the food and agriculture system, bringing together individuals with different methodologies, expertise, perspectives, philosophies, backgrounds, and networks offers an opportunity to generate new insights to achieving solutions to problems, and to produce broader impacts from the work of the collaboration.

Preliminary Observation #5: Inter-institutional collaboration can allow human, fiscal, and physical resources to go further and have a broader impact. Collaboration has the potential to reduce the duplication of effort and deploy scarce resources more efficiently when the roles of collaborators and the assets brought to the project are complementary. Partner institutions need not play identical roles, and collaborations can result in different kinds of outcomes, outputs, and impacts for the project and for the partners that are mutually beneficial.

SECTION 3: BARRIERS TO COLLABORATION AND IDEAS FOR OVERCOMING THEM

The charge to the Blue Ribbon Panel asks it to recommend actions to enhance the success of collaborative activities among institutions in the land-grant system. Identifying and overcoming barriers is an obvious need. A number of examples of potential barriers and means to overcome them can be found in the following list.

Preliminary Observation #6: Institutions use different approaches for approving funds to support faculty involvement in collaborations that may create varying expectations on the nature of collaborations and the role of participants. For example, some institutions use multistate funds to support salaries while others may only use them to support travel to participate in meetings. The rationale and implications for these differences needs to be explored further.

Preliminary Observation #7: Historical inequities have handicapped the ability of many 1890 and 1994 institutions to be full partners in collaborations with the 1862s. Different authorities guide the allocation of federal funds to 1862, 1890, and 1994 institutions respectively, with different requirements for state matching support of federal dollars, and different stipulations for the use of funds for collaboration. There is an urgent need to explore how to level the playing field for collaborative opportunities across differently resourced universities and ensure full and equitable participation among all collaborators.

Preliminary Observation #8: Faculty members may already be fully committed to other grants and teaching assignments, leaving insufficient or inadequate time and resources to support new collaborative projects. Younger faculty, in particular, may face strong disincentives for pursuing collaboration. Modifying evaluation criteria for promotion and tenure to ensure that participation in collaborations is appropriately recognized and rewarded has been a long-standing need. In 1890 and 1994 institutions, fewer faculty members are generally available to share teaching, research, and extension responsibilities. The potential to provide direct funding for those who have heavy teaching responsibilities to "buy themselves out" while involved in collaboration and providing "teaching postdocs" should be explored. Ensuring adequate start-up resources for newly hired faculty and investing in "continuation" resources for newly promoted faculty may also offer possible solutions.

Preliminary Observation #9: Land-grant system institutions have traditions emphasizing and rewarding competitive, rather than collaborative, research projects. Requirements to fund administrative and overhead costs to all institutions participating in a collaboration may also reduce incentives to share funds among multiple institutions. This situation is counter to the impression of the land-grant system as a unified enterprise. In exploring solutions, it is notable that the National Science Foundation (NSF) supports collaborative projects in which each PI gets their own budget for their work on the project, thus minimizing bureaucracy.

Preliminary Observation #10: A lack of information about the distribution of expertise at institutions across the land-grant system or of other available assets may hinder the ability to identify suitable partners for collaboration. Collaboration organizers may not be aware of the potential for external public or private partners with assets (data platforms, technologies, funds, networks, etc.) to participate and support land-grant collaborations. Collaboration could be encouraged by using NIFA to convene a workshop or series of workshops highlighting individual investigator projects across the agency that are doing work in the same or similar areas.

Preliminary Observation #11: The time available for planning collaborations properly is sometimes inadequate. Planning is the methodical process for envisaging collaborative activities and taking steps to provide what is required for their successful implementation. Opportunities to obtain planning grants could be expanded with criteria to achieve multiple goals. For example, NIFA could fund preliminary studies or preproposals that require three institutional types (e.g., large 1862, small 1862, 1890, minority serving institutions, 1994s, etc.). An evaluation criterion for competitively awarded funding that gives inter-institutional collaboration sufficient weight by funders could make the effort more worthwhile to would-be grantees. The challenge to applicants would be to include a diversity of disciplines, people, perspectives, and backgrounds. An open question is whether such planning awards should be focused on topics or projects of a specific scope.

Preliminary Observation #12: Leading collaborations requires team building, emotional intelligence, and project-management capabilities, for which many faculty are unprepared and untrained while administrative supports may not be available to assist. Institutions might proactively identify faculty leaders and prepare them to lead collaborations. Returning indirect costs to cover additional administrative support might assist new leaders and improve the success of multi-institutional collaborations.

Preliminary Observation #13: Institutions have different administrative procedures and policies for proposals, agreements, intellectual property, reporting, and mechanisms for handling funds that may create time lags, paperwork burdens, and opportunity costs that discourage collaboration. Institutional leaders can reduce the activation energy needed for inter-institutional collaboration by addressing these issues. One way to achieve this might be through the development of broad agreements between institutions that describe a common vision of needs, aspirations, perceived benefits of collaboration, and mechanisms to support participation. Establishing such agreements would require negotiation and a commitment of time, but the results could produce trust, awareness, and insight into the partner institutions' culture, norms, breadth of expertise, and administrative practices. Such agreements send a signal to faculty that there is buy-in from leadership to support collaboration.

SECTION 4: AMPLIFYING AND COMMUNICATING THE IMPACTS AND OUTCOMES OF COLLABORATION

Impact is the powerful effect that an activity or project, especially something new, can have on a situation, person, or policy. Amplifying impact means to intensify or expand that effect. The panel observed the potential for amplifying outcomes on several levels.

Preliminary Observation #14: The size and complexity of a collaboration should be expected to change over time and take on new goals and partners. The goals of a project would dictate whether a national scale, multifaceted, multisector effort must be built from the start of a collaboration, or whether a pilot phase or regional effort could be useful in providing proof of concept. Alternatively, a project may be organized with multiple phases and with different kinds of activities planned for early versus more mature stages of a partnership. It may be possible to broaden impact by building on top of an existing collaboration by, for example, using competitive extension funds to bring knowledge or practices identified in the first phase of a collaboration to a broader audience of users, or simply to support the dissemination of information in new formats and tools.

Preliminary Observation #15: Some key issues and questions in agriculture may require large, multidisciplinary collaborations and sustained research over time in multiple locations. NSF provides sustained long-term funding through its LTER and LTAR sites that supplements individual projects, and the National Institutes of Health supports Cancer Centers and numerous other multicomponent projects. The potential for this type of mechanism to create longer and more sustained success in agricultural research needs exploration. Funding from agencies like NSF, NIH, NASA, DOE, and EPA might be leveraged to broaden collaboration across the country.

Preliminary Observation #16: Collaborations lead to a diversity of outcomes, including some for which impacts are not easily recognizable or require more time to achieve. Greater thought on how to measure and communicate the value of those outcomes would increase their visibility. Anticipating expected outcomes and how to communicate them, defining what success looks like, and creating metrics for impacts could increase the effectiveness of communication efforts. To strengthen the ability to capture and communicate impacts, NIFA might provide additional support for post-collaboration assessments.

Preliminary Observation #17: Stakeholders, including producers, policy makers, and the public, are more likely to recognize the impacts and benefits of collaborative activities among land-grant institutions if communication strategies are tailored to them. Documenting the economic, environmental, and societal impacts and benefits of collaboration could raise the profile of the land-grant system to stakeholders. Infographics and other visualization methods, such as graphs and graphics, and using different media types, formats, and information portals for disseminating information would make communication more effective.

NIFA has a communication unit that collects, writes and transmits to the public the impacts of work that it supports (https://nifa.usda.gov/impacts). The NIFA staff also provide workshops on recognizing and effectively communicating impacts. NIFA's "Share Your Science" campaign (https://nifa.usda.gov/share-your-science) is designed to highlight research outcomes and accomplishments on a national level. It is aimed at spotlighting the achievements being made by NIFA's partners in addressing societal challenges, such as increasing food security and decreasing hunger, and addressing climate change, food safety, childhood obesity, and sustainable energy. This effort is critically dependent on submissions from the LGU system and USDA communications staff and could benefit greatly from ready access to institutional impact stories.

Another successful model is the Multistate Research Fund Impact Project (https://www.mrfimpacts.org) created in 2010 by the Experiment Station Section on Organization and Policy (ESCOP). It works with multistate committees to develop impact statements as their projects terminate. Project staff provide workshops on recognizing and communicating impacts for project participants, faculty, and staff. ESCOP maintains a social media presence and is also linked to the USDA Communications Office. A complete listing of impact statements is available on the project website.

Appendix D

Responses from Stakeholders to Preliminary Observations

The Panel's preliminary observations (see Appendix C) were posted online from April 14 through May 10, 2022, along with an online portal that presented observations sequentially in abbreviated form with the option to comment or skip. A total of 84 individuals left comments. Respondents were given optional questions about the type of organization for which they work (see Table D-1), job titles (see Table D-2), and location (see Table D-3). Table D-4 is a snapshot of agreement or disagreement and short topics mentioned by respondents. Complete responses are available from the project's public access file at www.nationalacademies.org.

TABLE D-1 Type of Organization with Which Respondents Are Affiliated

Type of Organization	Number of Respondents
1862 land-grant institution	43
1890 land-grant institution	9
1994 land-grant institution	7
Regional Association of State Agency Experiment Station Directors	2
Regional Association of Extension Directors	2
Association of 1890s Research Directors	1
First-Americans Land-Grant Consortium	1
Extension Committee on Organization and Policy	1
Experiment Station Committee on Organization and Policy	1
State Science Teachers Association	1
Extension Foundation	1
Federal agency	4
State agency	3
Agricultural producer	1
Private company	1
Total	78

TABLE D-2 Job Titles of Respondents

Job Title	Number of Respondents
Experiment station director	2
Extension administrator	9
Faculty (tenured professor or associate professor)	18
Faculty (non-tenure track)	5
Dean	1
Department chair	5
Lab technician	1
Land-grant director/coordinator	4

Land-grant outreach specialist	1
Library director	1
PhD candidate	1
Recruitment/student support	1
Sr. extension associate	1
Unit director	1
Vice president for research	2
Total	53

TABLE D-3 Summary of Respondents' Locations

Location	**Number of Respondents**
Midwest	22
Northeast	17
South	17
West	19
Other (Samoa)	1
Total	76

TABLE D-4 Topics Mentioned in Responses to the Observations

Observations and Topics Mentioned in Responses	**Number of Responses**
Obs. 1: There are already many ongoing inter-institutional collaborations in the land-grant (LG) system. Topics mentioned: - Examples of successful collaborative projects/programs - Uneven participation in collaborations - Forced participation in large-scale projects - Collaboration will not occur without first building a relationship - Opportunities to bridge gaps - Meaningful participation requires equity - Individuals will seek out those with whom they have had prior success - Must include non-LGs in the perspective	38 Agree: 21 Disagree: 17
Obs. 2: Inter-institutional collaborations do not routinely engage faculty from across the full range of LG colleges and universities. Topics mentioned: - 1862s just check box when seeking collaboration with 1890s or 1994s - Historical and current inequities - Regional rural development - Cluster collaborations - Function of the research goals; could be tailored to 1890 and 1994 focus - Funding excludes some institutions	34 Agree: 23 Disagree: 11
Obs. 3: The nature of key questions for food and agricultural science are evolving, and the scientific approaches to address them are increasingly at the convergence of multiple disciplines and involve data collection across multiple geographies.	0

continued

TABLE D-4 Continued

Observations and Topics Mentioned in Responses	Number of Responses
Obs. 4: All of us are smarter than one of us: diversity promotes novel ideas. Topics mentioned: - Diversity, equity, and inclusion need attention - Indigenous knowledge is important - This is obvious and will happen when funding is provided	28 Agree: 25 Disagree: 3
Obs. 5: Collaboration can allow human, fiscal, and physical resources to go further and have broader impact. Topics mentioned: - Yes, but mostly in theory - Indirect costs eat up budgets - Management of bureaucracy comes with a cost - LGs need to embrace National Science Foundation model of broader impacts - Systems between 1862s and 1994s too different and lead to frustration	24 Agree: 11 Disagree: 13
Obs. 6: Institutions use different approaches for approving funds to support faculty involvement in collaborations that may create varying expectations on the nature of collaborations and the role of participants. Topics mentioned: - A capacity issue, as teaching is the greatest need before collaboration. - Underfunding by state is a common reason multistate funds are used for base services - May be on reason some do not seem to want to work with an HBCU - Each member should know resource allocation and ability to contribute - Research funding needed for two-way collaborations	8 Agree: 6 Disagree: 2
Obs. 7: Historical inequities have handicapped the ability of 1890 and 1994 institutions to be full partners in collaboration. Topics mentioned: - More trust building needed - Competing demands may be an unsurmountable impediment to collaboration - Historical and current inequities continue - Matching funds not received - Support needed for more robust fiscal management	27 Agree: 23 Disagree: 4
Obs. 8: Faculty members may be fully committed to teaching and other assignments, leaving insufficient time and resources to support collaborative projects. Topics mentioned: - That is the nature of the position; everyone sets their own priorities - Overcommitment is true for faculty at all land-grant colleges and universities - Early-career faculty must be cautious when pursuing collaboration	28 Agree: 20 Disagree: 8
Obs. 9: LG colleges and universities emphasize and reward competitive, rather than collaborative, research projects. Topics mentioned: - Old mindset/nationalistic attitude persists - Would not be the case if funds for collaboration would be adequate - Competitive grants bring in more money - We spend too much time seeking resources and not working on projects	30 Agree: 17 Disagree: 13

Obs. 10: Lack of information on the distribution of expertise across the LG system may hinder the ability to find suitable partners for collaboration. Topics mentioned: - Principal investigator (PI) to PI workshops best way to find partners - Funding agencies need to do a better job teaching people how to query databases for partners - Interpersonal networks and social media are needed - Fund travel for faculty to attend meetings	20 Agree: 11 Disagree: 9
Obs. 11: Time for planning collaborations is sometimes inadequate. Topics mentioned: - National Institute of Food and Agriculture Specialty Crop Initiatives use planning grants - Big institutions have grant writers - Need to start planning a week after the Request for Applications comes out - Relationships have to be built before collaborations can happen - Never enough time to plan	24 Agree: 16 Disagree: 8
Obs. 12: Leading collaborations requires team building, emotional intelligence, and project management skills that many faculty lack. Topics mentioned: - Being a collaborator is different than being a principal investigator - Training is needed but worry about useless training - Offer sabbaticals to undertake leadership of team - Training is needed - Full-time manager who is not a faculty member is ideal	23 Agree: 17 Disagree: 6
Obs. 13: Differing administrative procedures and policies may create time lags, paperwork burdens, and opportunity costs that discourage collaboration. Topics mentioned: - Need to allow funds to be awarded then negotiate the agreement between institutions; otherwise, a waste of time if grant is not funded - Mandated policy on indirect costs is needed - Standardization is not possible - LGs should be working this out ahead of time	21 Agree: 15 Disagree: 6
Obs. 14: The size and complexity of a collaboration should be expected to change over time and take on new goals and partners. Topics mentioned: - Partner with team scientists - Yes, but sometimes it takes a long time to fulfill an objective - Expand grant timelines - Have to deal with natural turnover of leadership	12 Agree: 12 Disagree: 0
Obs. 15: Some key issues in agriculture may require large multidisciplinary collaborations and sustained research over time in multiple locations. Topics mentioned: - Yes, but this could separate the have and have nots further - Yes, but not possible on 3–5 year grants - Examples of research projects: climate, biosecurity, soil management - Need an infusion of money and cross-sectoral participation - We need convergence	20 Agree: 16 Disagree: 4

continued

TABLE D-4 Continued

Observations and Topics Mentioned in Responses	Number of Responses
Obs. 16: Collaborations lead to diverse outcomes, including some which require more time to achieve. Topics mentioned: - True, so engagement of evaluators to study the science of team science is needed - True, so short-, middle-, and long-term goals are needed	15 Agree: 11 Disagree: 4
Obs. 17: Producers, policy makers, and the public are more likely to recognize the impacts and benefits of collaborations if communication strategies are tailored to them. Topics mentioned: - Do not tailor it to them, engage them - Need a major marketing effort to communicate - Researchers need to be held accountable	21 Agree: 13 Disagree: 8